Praise for *The Thinking Beekeeper*

No matter the box you keep your bees in, if you are a new beekeeper you need solid, practical and most of all accurate information to get started. You'll find that here. And if you are keeping your bees in a top bar hive, you'll find information you need here that's not available anywhere else. Both you and your bees will benefit from Christy's approach, advice and philosophy.

— Kim Flottum, editor *Bee Culture* Magazine

Christy Hemenway's *The Thinking Beekeeper* is a very nice book. It provides a blend of the author's philosophy, ranting (about the use of chemicals in beekeeping), and clear practical advice about honey bee culture, especially regarding top bar hives. There has been little written about the specifics of raising honey bees in top bar hives. The bees are the same of course, but the top bar hive is quite different from the traditional Langstroth Hive. At the University of Maine we had plenty of questions when we first embarked upon the use of the top bar hive. NOW there is a good guide. Not only is the book informative, being accessible to all with its clear concise prose and liberal use of photos and data tables, but in addition it is enjoyable to read.

— Dr. Frank Drummond, pollination ecologist, University of Maine

Christy's passion shines through in this delightful book, which I'm sure will inspire many people to take up top bar beekeeping. I am particularly pleased to see that she has developed her own style, while staying true to the principles of simplicity and minimal interference with the lives of the bees. Having watched Christy's progress so far, I'm sure that Gold Star Honeybees has a great future!

— Phil Chandler, author, *The Barefoot Beekeeper*

It is great to see that top bar beekeeping is alive and well in Maine and that Christy Hemenway is passionate about her top bar hives. The top-bar hive is coming!

— Les Crowder, coauthor, *Top-Bar Beekeeping*

Whether you're looking for another argument for keeping your own bees or are already convinced, *The Thinking Beekeeper* is an excellent resource. Christy knows her stuff and shares her experience and passion on every page.

— Roger Doiron, Founder, Kitchen Gardeners International

The Thinking Beekeeper is a unique and exceptional resource for the beginning beekeeper. It will enable the novice to make a successful start in the craft and as he/she progresses all those instructions offer the opportunity to object to something Christy recommends. And that ladies and gentlemen is the badge of an independent practitioner and mature thinking beekeeper.

— Marty Hardison, top bar beekeeper,
educator and international developmental beekeeping consultant

the *thinking* Beekeeper

Today, more than ever before, our society is seeking ways to live more conscientiously. To help bring you the very best inspiration and information about greener, more sustainable lifestyles, *Mother Earth News* is recommending select books from New Society Publishers. For more than 30 years, *Mother Earth News* has been North America's "Original Guide to Living Wisely," creating books and magazines for people with a passion for self-reliance and a desire to live in harmony with nature. Across the countryside and in our cities, New Society Publishers and *Mother Earth News* are leading the way to a wiser, more sustainable world. For more information, please visit MotherEarthNews.com.

Join the Conversation

Visit our online book club at www.newsociety.com to share your thoughts about *The Thinking Beekeeper*. Exchange thoughts with other readers, post questions for the author, respond to one of the sample questions or start your own discussion topic. See you there!

A Guide to
NATURAL BEEKEEPING
in **TOP BAR HIVES**

The
thinking
beekeeper

CHRISTY HEMENWAY

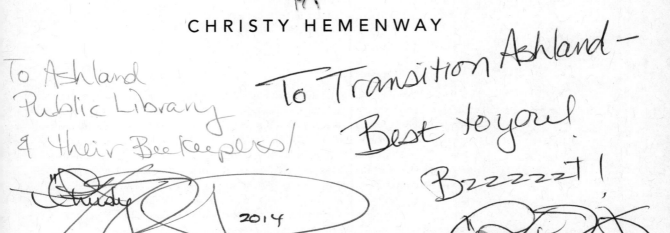

To Ashland
Public Library
& their Beekeepers!

2014

To Transition Ashland —
Best to you!
Bzzzzzt!

new society
PUBLISHERS

Cover design by Diane McIntosh.
Honeycomb image © iStock (rvbox); Bees © iStock (antagain);
Handmade paper background © iStock (Kim258); Top bar hive © Christy Hemenway
Back cover image of Queen Bee copyright Tony Jadczak/Maine's State Apiarist.

Printed in Canada. First printing January 2013.
Paperback ISBN: 978-0-86571-720-6
eISBN: 978-1-55092-511-1

Inquiries regarding requests to reprint all or part of *The Thinking Beekeeper*
should be addressed to New Society Publishers at the address below.

To order directly from the publishers,
please call toll-free (North America) 1-800-567-6772,
or order online at www.newsociety.com

Any other inquiries can be directed by mail to

New Society Publishers
P.O. Box 189, Gabriola Island, BC V0R 1X0, Canada
(250) 247-9737

LIBRARY AND ARCHIVES CANADA CATALOGUING IN PUBLICATION

Hemenway, Christy
The thinking beekeeper : a guide to natural beekeeping
in top bar hives / Christy Hemenway.

Includes index.
ISBN 978-0-86571-720-6

1. Bee culture. 2. Bees. I. Title.

SF523.H46 2012   638'.1   C2012-907149-8

New Society Publishers' mission is to publish books that contribute in fundamental ways to building an ecologically sustainable and just society, and to do so with the least possible impact on the environment, in a manner that models this vision. We are committed to doing this not just through education, but through action. The interior pages of our bound books are printed on Forest Stewardship Council®-registered acid-free paper that is 100% **post-consumer recycled** (100% old growth forest-free), processed chlorine free, and printed with vegetable-based, low-VOC inks, with covers produced using FSC®-registered stock. New Society also works to reduce its carbon footprint, and purchases carbon offsets based on an annual audit to ensure a carbon neutral footprint. For further information, or to browse our full list of books and purchase securely, visit our website at www.newsociety.com

Contents

Acknowledgments

Lots of folks get dragged into the mix of life that then manifests as a book, it seems. Here is a short and assuredly incomplete list of folks who had a great deal to do with the eventual creation of this book.

Phil Chandler: for the original inspiration and continuing support and friendship

Michael Bush: for his practical outlook, no bull attitude and his friendship as well

Kathy Keatley Garvey: for her award-winning sting photo

Kim Flottum and Kathy Summers: for real answers to real questions, being open-minded, chemical-free and great hosts

John and Ruth Seaborn: for their wonderful, treatment-free bees

Suzanne Brewer: for her friendship and for producing the professional construction drawings that you can now purchase in Gold Star Honeybees®' DIY kits.

Jim Fowler: for his help, continuing support and for taking pictures when my hands were full

Knox Lincoln County Beekeepers: my bee school alma mater

Gunther Hauk: for his dedication to and his outlook on beekeeping

All the Northeast Treatment Free (NETF) folks

Dee Lusby: for being such an advocate of chemical-free beekeeping

Dennis Murrell: for his blog

The folks behind the movies *Vanishing of the Bees* and *Queen of the Sun*

Jay Evans at the Beltsville Bee Lab

Maryann Frazier, Senior Extension Association at Penn State University

Dennis vanEngelsdorp: for his ability to get beekeepers, researchers and cooperative extension to all pull together in the same direction.

Rowan Jacobsen: for writing *Fruitless Fall*

David and Linda Hackenberg

All the suppliers and vendors that have played a part in producing Gold Star Top Bar Hives since the kits were launched.

And finally, Pam Tetley for her editing help early on in the process before I knew anything at all about how to put a book together and Betsy Nuse, who was there with her editing expertise to catch this book as it was finally born.

Introduction

Never doubt that a small group of thoughtful, committed citizens
can change the world; indeed, it's the only thing that ever has.

« Margaret Mead »

Welcome to the world according to Christy Hemenway...

This book is a result of my first year as a backyard beekeeper — with two conventional, Langstroth, square-box hives containing sheets of wax foundation — and my switch to top bar beekeeping.

This is a how-to/why-to book. It is the amalgam of my own personal beekeeping experiences with the writings, the experience, the research and the bee stories of many amazing people — beekeepers, farmers, gardeners, activists, researchers, authors — whom I've encountered since I had those first two hives.

This book does its best to ease the bewilderment that I remember feeling when I decided I wanted to start beekeeping — and then discovered that if you asked 10 beekeepers a question, you were sure to receive 11 answers, many of them expressed vehemently, and in "no uncertain terms." As a novice, you hope for just one answer...the right answer — only to discover that there are a hundred ways to keep bees. It's confusing! I try to make this research phase easier for you with the how-to parts of the book.

In the why-to parts of the book, I address the paradigm shift that I've seen gaining momentum since Colony Collapse Disorder made its debut in 2006, and my subsequent founding of Gold Star Honeybees® in 2007.

It's pretty clear that the crisis of Colony Collapse Disorder in the beekeeping world is a symptom of wider problems in our environment and food systems and cannot be remedied just by those of us who keep bees.

As we became aware of the many connections between beekeeping, our broken food system, governmental corruption and our own health and well-being, thinking people started to wonder…how did we get to this point? How do we make the changes we need to make to correct these problems? Changes that are a matter of life and death — to us and to our children…

In the US we've grown tired of expecting that the government will take charge, behave responsibly and do the right thing. But — we don't have to wait for the government to make the right move! We can make the needed changes, and the government — well, they can catch up. We can insist on organic food, and we can shop at the farmer's market, and we can choose never to put anything in a beehive but bees…these are all viable options, and thinking people are doing them, and they are making a difference.

That's why I believe that the paradigm truly has begun to shift. In fact, I think we're close to the tipping point. And I also believe that we don't have to find a cure — a new treatment, pesticide or antibiotic — for Colony Collapse Disorder — we just have to quit causing it.

For those of you who get it that honeybees are part of a huge, important, delicate and complex natural system — and who think that you would like to do your own part for that system and for them — this book is for you.

In my mind, you will always be iconoclasts, rebels, renegades…in other words…

Thinking Beekeepers.

What you do matters. Never doubt it.

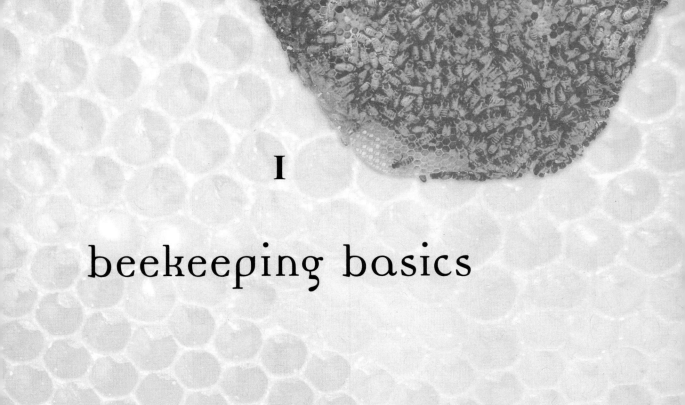

I

beekeeping basics

CHAPTER 1

How Did We Get Here From There?

Humankind's interaction with bees spans many thousands of years. But the relationship has not always been as one-sided as it is today.

Ancient civilizations were *honey hunters* — collecting honey from beehives discovered in the wild. Often this included physically destroying the hive in order to gather the honey.

At some point, humans began attempting to domesticate the honeybee. While the idea of actually taming a honeybee is a bit cheeky, people did manage to convince bees to live in containers or cavities of our choosing, in locations that we selected. These containers became known as beehives, and they included such things as hollow logs, pottery vessels, wooden boxes and woven straw baskets.

The ancient Egyptians were probably the first culture to maintain bees in artificial hives. They floated barges carrying clay hives up and down the Nile River where flowers were plentiful. The bees would forage along the river during the day, and then the barges were drifted down the river at night following the source of food as new flowers bloomed through the season. It is said that archeologists found sealed pots of honey that were still edible in the tomb of King Tut (1341 BC–1323 BC).

Thirty intact beehives (circa 30 BC) were found in the ruins of the Jewish city of Rehov. The hives were made of straw and unbaked clay and arranged

The original top bar hive.

Credit: John Caldiera. "American Beekeeping History—The Bee Hive." *John's Beekeeping Notebook.* [online]. [cited July 26, 2012]. outdoorplace.org/beekeeping/history1.htm.

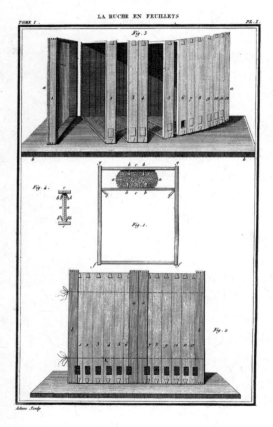

Francis Huber's leaf hive.

Credit: Plate of leaf hive from Huber's *New Observations Upon Bees*, X-Star Publishing, Copyright 2012 by Michael Bush.

in rows. This places beekeeping and a fairly advanced honey industry at the time of the Bible, about 3,000 years ago.

The first top bar hive — a movable comb hive — is said to have been in use in Greece in the 1600s AD. The bars were placed across a container, like a basket, and spaced so that the bees drew their combs in such a way that they could be safely lifted out and inspected. Francis Huber is credited with developing the first beehive with movable frames in Switzerland in 1789. Known as the Leaf Hive, the frames of the hive were hinged at the back and could be turned like the pages (leaves) of a book. It was Huber's hive that led the Reverend Lorenzo Lorraine Langstroth to feel confident that it would be possible to build a hive that would allow for the inspection of the hive without "enraging the bees."[1]

In the mid-1850s, Reverend Langstroth designed a hive with removable, self-spacing frames that worked in correlation with the concept of *bee space* — the ⅜ of an inch that bees need in order to move between the combs. Where there are areas in the hive that are larger than ⅜ of an inch the bees will fill those spaces with comb, and areas less than ⅜ of an inch will be closed up with *propolis*.

Controlling the spaces between the frames made manipulating the contents of the hive a simpler matter for the beekeeper — leading to greater so-called efficiency in beekeeping.

In Langstroth's own words:

> …the chief peculiarity in my hives…was the facility with which these bars could be removed without enraging the bees…. I found myself

able…to dispense entirely with natural swarming, and yet to multiply colonies with much greater rapidity and certainty than by the common methods. I could, in a short time, strengthen my feeble colonies, and furnish those which had lost their Queen with the a means of obtaining another. If I suspected that any thing was wrong with a hive, I could quickly ascertain its true condition, by making a thorough examination of every part…[2]

The original Langstroth Hive.
Credit: John Caldiera. "American Beekeeping History—The Bee Hive." *John's Beekeeping Notebook.* [online]. [cited July 26, 2012]. outdoorplace.org/beekeeping/history1.htm.

Langstroth's frames supported the comb on four sides, so there was less risk of breaking the comb when it was being handled and inspected. By the end of the 1800s most North American beekeepers were using some variation of the Langstroth hive.

In the early 1920s, Rudolf Steiner gave a series of lectures entitled *Bees*.[3] He was very concerned about the level of manipulation and mechanization that was occurring in the beekeeping world, even though it tended to make beekeeping far simpler as an industry. His concerns, even then, included:

- the use of ready-made combs, i.e. foundation
- the manipulation of queen bees
- using worker eggs to manufacture queen bees
- striving to thwart the natural swarming/reproductive impulse of bees
- monocrop agriculture
- moving hives to pollinate crops
- the use of chemical fertilizers
- the use of pesticides.

Steiner viewed all these things as meddling with natural systems, and as posing risks and detriments to the health and sustainability of both bees and agriculture.

One lecture in 1923 is notorious for a conversation that occurred between Steiner and a beekeeper by the name of Mr. Muller. In this conversation Steiner stated that the artificial queen breeding methods that were being developed at the time would cause serious bee collapse within one hundred years. Mr. Muller did not agree. All indications seem to say that Rudolf Steiner was spot on with his concerns, as well as with the timing of his prediction.

Forward, into the Future of Food

Significant events in recent history have influenced the world of bees and agriculture. Here are some of them:

The Products of War

The use of chemical fertilizers and pesticides arose from the aftermath of two World Wars. Fritz Haber, the German Jewish chemist who discovered the means of synthesizing nitrogen, forever changed the fertility of the planet. Haber also developed the cyanide-based pesticide, Zyklon A, which was later adapted as Zyklon B, the gas used by the Nazis in concentration camp gas chambers.

"As the Indian farmer activist Vandana Shiva says in her speeches, 'We're still eating the leftovers of World War II.'"[4]

Big Agriculture

In 1971, US President Richard Nixon appointed Earl "Rusty" Butz as the Secretary of Agriculture. With Earl's aggressive encouragement to "get big or get out," US farmers bought heavily into the ideas that big was better — that monoculture was a more efficient method of growing food — and that the chemically induced increase in yield that monoculture created (which was fortified and protected by the products of war — chemical fertilizers and pesticides) would one day feed the world.

Big Ag promoted monoculture farming practices as better and more efficient. But in reality, monoculture practices destroy the critical balance between soil, water, livestock, crops and pests — the very features that keep agriculture alive. Today any system that does so much damage to itself instantly earns the moniker "unsustainable." The damaging effects of Earl Butz's policies are still being felt today.

Systemic Neonicotinoid Pesticides

After years of traditional pesticides, a completely new class of pesticides—systemic neonicotinoids—came on the scene in the 1990s. These were not without their supposed advantages; after all, they were intended to reduce the amount of toxic chemicals being sprayed on plants, and this they did.

And that sounds like a good thing…instead of being sprayed on the developing plant, a systemic pesticide is painted on the seed of the plant. As the plant grows, the pesticide is in the very tissue of the plant, not just on the plant—in other words, you can't just wash it off. The use of systemic neonicotinoids also means that their poisons are in the nectar and the pollen of the plant—the part that constitutes bee food.

Outmoded Testing Methods

The methods currently in use to test this new, insidious type of pesticide have proven to be woefully inadequate for measuring the danger to honeybees. The current methods don't take into account the possibility of delayed responses to the pesticides, or the concept of *sublethal effects*—effects that damage but don't necessarily kill the honeybee. Any effect is relevant when one considers the interconnectedness of our food system and these important pollinators.

Genetically Modified Organisms

Bees and beekeepers now have to contend with another worrisome thing—genetically modified organisms (GMOs). The audacity of patenting a plant completely astounds me—along with the frightening implications of creating a plant that has been genetically altered in order to withstand pests and pesticides. The methods used to genetically alter plants could have deep implications on our health and on the future of the planet.

A battle currently being fought in Washington concerns the legal requirement to label GMO food, and I encourage thinking people to educate themselves on this important topic as it evolves.

Our Broken Food System

Michael Pollan, in *The Omnivore's Dilemma* and in *The Botany of Desire*, eloquently illuminates the fundamental brokenness of our food system.[5] He defines the 1,500-mile salad, does the math on the grossly inefficient use of fossil fuels in producing food, discusses the wacky sex life of corn and describes in vivid detail the effect that government-subsidized overproduction of this one mutant grass has had on the US farming economy—even though it does not require honeybees for pollination!

And now, Back to the Bees

In late 2006, news broke of a frightening problem happening with honeybees—a small insect so completely crucial to our food system that our lives literally depend upon its existence.

This problem with bees was named *Colony Collapse Disorder*, and it quickly became a major focus of the bee research community. It garnered a snazzy abbbreviation—CCD—and was soon a major buzzword in the media.

The primary symptom of CCD is frighteningly strange. Let me try to make it clear just exactly how strange it is.

First you must understand that bees sting in order to defend two treasures: their *brood* (baby bees) and their food (honey). And since when a honeybee stings, she dies—obviously to the bees, these two things are of life-or-death importance.

But when a colony collapses, the adult bees simply disappear. What the beekeeper finds is a hive containing brood and food, but no adult bees. The bees have abandoned the two things most important to them.

This also means that there are no dead bees—no bodies to study. They're just gone, having left behind all those babies and all that honey. It's eerie and almost too weird to contemplate.

In the years that have elapsed since 2006, bee researchers have gradually concluded that the cause of CCD cannot be pinned on any one single thing—one pesticide, one fungus, one virus, one parasite—but that CCD is caused by combinations of stressors breaking down the bees' natural systems. CCD is truly an indication that the bees have reached the limit of their ability to withstand the stress of the manipulations and mechanization that they've been subjected to.

For industrial beekeepers, especially large-scale migratory pollinators, Colony Collapse Disorder has been a devastating fiscal tragedy, not to be wished on anyone. And I have never, ever met a beekeeper — commercial, backyard or otherwise, who did not love their bees — so there is personal heartbreak as well in every vanished colony.

But on a different note, it could be that CCD is an opportunity for eaters and beekeepers everywhere to awake from a strange and mind-numbing slumber…and for us all to realize that

- big is not necessarily better.
- faster is not necessarily a good thing.
- more is not necessarily the goal.

And as bad as all this doom and gloom is, doesn't it also impart a sense of hope?

It does for me — because CCD points out a very, very important thing…

We don't have to find a cure for CCD — we just have to stop causing it!

And we can do that by respecting and working with the natural systems that are part of our food system, and at work inside the beehive.

When one tugs at a single thing in nature,
one finds it attached to the rest of the world.

« John Muir »

CHAPTER 2

It's All About the Wax

Until man duplicates a blade of grass,
nature can laugh at his so-called scientific knowledge.

« THOMAS EDISON »

Honeycomb

Honeycomb or comb — with its signature hexagons — is the beeswax structure that makes up the heart and skeleton of a honeybee colony. The hexagon shape itself is a marvel of natural engineering — it is the most efficient shape available. The six matching and touching sides of each cell use the least amount of material to create, while at the same time producing a very strong structure — and the hexagons fit together perfectly, leaving no wasted space between.

Comb is created by the bees and made from *beeswax*. Tiny white ovals of beeswax are secreted from the wax glands in a young honeybee's abdomen and shaped by the bees into sheets or panels of interconnected hexagons.

All of the natural processes of the hive — everything that bees do — happens in or on their comb. Bees raise their young (known as *brood*) in *brood comb* cells; they store their food in *honeycomb* cells, and they arrange the combs inside their hive according to the needs of the colony. Comb is intrinsic to the honeybees' existence.

Honeybees are known as *cavity nesters*, meaning they build their nests, consisting of multiple sheets or panels of comb, inside a sheltering space, or cavity. The cavity might be a hollow tree or branch, a hollow space in a cliff, a space inside the walls of a building, the interior of a beehive or other similar cavity.

The bees create the comb and situate it throughout the hive according to their needs — the raising of female *worker bees* (which requires a specifically sized cell), the raising of male *drone bees* (requiring a differently sized cell) and the storage of the pollen and honey that make up their supply of food.

A natural beehive filled with natural beeswax honeycomb should invoke wonder at the magic at work there. If you accept that honeybees have existed on earth for 65,000,000 years, give or take a few millennia, then it would seem that we owe a certain amount of respect to these industrious creatures and their innate knowledge of what they need. The natural systems at work inside a hive should be protected, not thwarted.

Natural beeswax—the heart and skeleton of the bee colony.
Credit: Jim Fowler.

Frames

In the 1850s the Reverend L. L. Langstroth applied for and received a patent for what we in the US now consider the typical square-box, conventional beehive. Langstroth's hive utilized *Hoffman frames* — self-spacing frames which help to maintain bee space between the combs and can easily be removed from the hive.

Bee space is the amount of space required for the bees to walk between their combs. Langstroth capitalized on the realization that if a hive and frames were constructed so that the space between the frames was a consistent ⅜ of an inch, then the bees were less likely to fill that space with *burr comb* (irregular comb not built within the frames) or try to fill it in completely with *propolis* (the strong, glue-like substance made by the bees from the resins of trees). The bees' ability and tendency to seal gaps in their hive with burr comb and propolis can be frustrating for the beekeeper, since it requires a great deal of work to keep the parts of the hive free and movable.

So in some ways, frames simplified life for the beekeeper. A *movable comb* hive allows for the early discovery of pests, diseases or queen problems, as the individual frames of comb can be removed and viewed. This type of hive also permits the removal of the frames of honeycomb for the extraction of the honey and subsequent return of the comb to the hive.

Since the frames provide a supporting structure around the comb, less caution is required when inspecting or *working the hive*. Frames can be handled fairly unceremoniously — beekeepers often set them on the ground, bump them against things (including each other) and lean them against the hive — and the comb supported inside a frame can withstand this sort of treatment.

In other ways, the use of frames complicated things for the beekeeper, as frames are time-consuming and laborious to put together, and their assembly occupies many a beekeeper during the winter months in anticipation of beekeeping season.

So frames in and of themselves are not necessarily detrimental to bees... and in fact they're sort of handy for beekeepers.

A frame from a Langstroth hive.
Credit: Christy Hemenway.

Foundation

Foundation is a sheet of wax or plastic with the outlines of hexagons embossed on its surface. The bees then *draw* their honeycomb based on this outline. Some foundation, known as *fully drawn foundation,* has the hexagonal cells completely constructed (drawn out).

Foundation is meant to serve as a guide for the bees' honeycomb — that special, natural structure so essential to the workings of the bee colony.

Why? Don't the bees know how to make their own honeycomb? Haven't they been doing just that for millions of years?

Good question. Here's part of the story.

Straight Comb

When beekeepers go to inspect their hives, it quickly becomes obvious that the comb needs to be straight so that it can be removed easily from the hive, inspected and then reinserted. So one reason for the development of foundation was to enforce the building of straight comb, contained within the frames of a conventional hive.

Head Start

Another intent behind foundation, and especially fully drawn foundation, is to offer the bees what is considered to be a head start — it provides cells which can be filled with brood or honey immediately.

The need for straight comb and the attempt to provide the bees with a head start were two beneficial driving factors behind the invention and use of foundation.

Foundation was never intended to be harmful to bees. No beekeeper begins with that in mind! In fact, the combination of movable frames and foundation went a long way toward savings entire hives of bees from being destroyed in order to harvest their honey.

But here's the rest of the story: Size matters!

Same Size

Part of the problem with foundation is that those raised hexagons embossed on the sheet of wax or plastic are all the same size. There is no evil master plan driving this fact, it's just that foundation is made by a machine. And machines know very little about the magical systems at work inside a hive of bees.

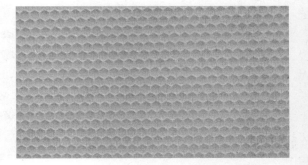

A sheet of standard wax foundation, with its one-size-fits-all cells.
Credit: Christy Hemenway.

But if you've ever seen natural honeycomb, drawn by bees, for bees — then you understand that *one size does not fit all*! Bees that are able to draw their own natural wax combs, without the use of foundation, make the cells the different sizes for different purposes: Worker bee brood is one size, drone bee brood is another, honey storage yet another…all these different sizes are very important to the inner workings of the colony.

Wrong Size

Now, not only are all the cells on a sheet of standard cell foundation the same size — but they are also the wrong size. The difference between *natural cell size* (4.9 mm) and *standard cell foundation* (5.4 mm) is only .5 mm. Just a little bit wrong — but significant when you consider the size of a honeybee.

Natural wax allows the bees to build the cell sizes they need.
Credit: Christy Hemenway.

Let me segue here and introduce a tiny, oval-shaped, rusty-brown parasite — a frighteningly effective disease vector and the bane of beekeepers everywhere — the *varroa mite*. Since the mid-1980s, beekeepers in North America have been wrestling with this ubiquitous parasite. Apparently we inherited varroa mites from Asian honeybees, which were able to live symbiotically with them. But when the European races of honeybees living in North America were exposed to the varroa mite, they were mostly defenseless.

That tiny difference between the cell size of natural beeswax comb and standard cell foundation — that extra half a millimeter — has made it easier for varroa mites to thrive and harder for honeybees to resist them. Enlarging the size of the cell even just that little bit extends the gestation period of the honeybee by approximately 24 hours. Part of that time — somewhere between 8 and 24 hours of it — occurs during the bee's larval stage.

The larval stage of the honeybee is the time during which the adult female varroa mite must enter the cell in order to be inside the cell after it is capped, and in order to breed. Once inside, the mite burrows into the brood food, beneath the larvae, extends its little snorkel-like breathing tubes and waits for the cell to be capped. Once the cell is capped the female varroa mite then emerges from beneath the larvae and proceeds to lay eggs.[1]

Extending the length of the larval stage of the developing bee increases the odds in favor of the varroa mite by allowing more time for her to enter the cell.

While no one set out to hurt the honeybee by using foundation, using it seems to have made both bees and beekeepers victims of the law of unintended consequences.

More Unintended Consequences

There are more ways beekeepers have denied the perfection of the natural systems at work in the superorganism that is a bee colony.

The two advertisements shown on pages 18 and 21 list what were perceived as the benefits of the new methods in beekeeping available with the advent of the Langstroth hive.

There is a benefit that the beekeeper be able to see "every inch of the hive," to know the status of the queen, the brood and the hive's food stores. It is also an improvement to not have to destroy the hive and the bees in order to harvest honey.

However, here are some other features of hives and foundation that demonstrate how easy it is to become careless with the manipulation of nature's systems. Some of the features listed may indeed be considered beneficial, but some are manipulative at such a deep level as to be pretty horrifying.

Preventing Natural Swarming

Swarming is the reproductive process of every bee colony. According to some internal clock that is all their own — driven by weather, the size and health of the colony, the resources currently available to them in their locale and the amount of space remaining in the cavity of their hive — bees will begin swarm preparations. The steps involved in this mysterious process include the raising of drone bees and the production of *swarm cells*, through which the existing queen helps to create her own replacement by laying eggs into vertical swarm cells in order to produce a new queen bee. Swarming is

REV. L. L. LANGSTROTHS'

Patent Movable Comb Bee Hive

THE subscriber having purchased the Patent Right for the above named Hive, to several counties in Ohio, and prepared himself for their manufacture, at his mill in Scioto township, Delaware county, Ohio, now offers it for sale. This hive is the result of the long and careful study and experience of an *extensive, intelligent and practical Bee-Keeper*, and in offering it to the public, the undersigned feels that he is *not* offering the mere contrivance of a scheming speculator, but an article of REAL, PRACTICAL, GENUINE WORTH, which will MEET THE DIFFICULTIES which have for hundreds of years thwarted the aim and hopes of the Bee Keeper. It is exceedingly simple in its construction and arrangement, while it gives the Apiarian perfect control, not only of his bees, but of the MOTH or any other enemy to their prosperity, he having free access, not only to every part of the hive, but EVERY INCH OF COMB IN IT, and that without the slightest danger or injury to either himself, the bees, honey comb, hive, or anything else.

To enumerate fully ALL its advantages would require too much space for an advertisement; but the following are some of them:

Destruction and prevention of injury by BEE MOTH.

Free access to every inch of the hive, and comb in it, without injury.

Knowledge (not conjecture) of the precise condition of the bees and comb at pleasure.

Swarming the bees, (at will,)—thus saving the necessity of watching for weeks.

Multiplying the number of swarms at pleasure.

Preventing natural swarming.

Strengthening weak or small swarms at any and all times.

Taking the honey in any form or quantity desired without destroying the bees

Selecting and taking the new and leaving the old honey at will.

Feeding the bees in cold or warm weather without exposure.

Protection to the bees in extreme cold weather.

Adaptation of the hive to the size of the swarm —large or small.

Preventing the accumulation of drones, which consume but do not gather honey.

Ascertaining when the queen is lost or dead, and supplying another.

In short there is no operation necessary to be performed which cannot be safely and easily done with this hive, and it opens to every intelligent man a new, easy, safe, and certain mode for successful bee culture.

The best proof of its superiority is the fact of its superceding all other hives, and retaining its supremacy wherever it has been used, yet it is so simple that all who see it are struck with wonder that it was no sooner discovered or invented. My agent (Mr G. W. Newlove) will be happy to shew the hive, with and without bees in it, to those who may call at my place. Price for hive and individual right $10.

RICHARD COLVIN,

5 miles west of Delaware on the S. M. & P R.R

N. B.—The Inventor of the above hive has published a most valuable work on ' Bee Culture," the result of 19 years arduous study and costly practical experiments which every bee keeper should read and study it may be had by those who wish it of my agent—Price $1.50.

Bees will be transferred from any hive into the Langstroth hive by Mr. Newlove, at my place free of charge   July 15, m3

Advertisement for Langstroth hives in the Marysville, OH Tribune, 1858.

Credit: Joe Waggle. "1858, Patent Movable Comb Bee Hive." Historical Honeybee Articles yahoo group. [online]. [cited August 27, 2012]. pets.groups.yahoo.com/group/HistoricalHoneybeeArticles/files/22%29%20Inventions%20/.

the bees' way of betting on the future and perpetuating their species. Bee-keepers have always seemed to want to control and mechanize the reproductive process of the honeybee, and conventional beekeepers are commonly taught to cut out and destroy any swarm cells they find during an inspection in order to prevent swarming. Not only is it nearly impossible to thwart the swarm impulse once it has begun — but it's likely that the destruction of swarm cells in the hive will leave the colony queenless — as the founding queen is the queen that departs with the swarm. Crushing the swarm cells the bees have made in preparation for reproducing is likely to leave them with no queen at all.

Multiplying the Number of Swarms at Pleasure

The modern equivalent of this is known as *splitting*. The beekeeper can divide the resources of one colony into two hives — and so long as the bees in the hive who wound up without a queen are able to produce a new queen bee, or the beekeeper provides an artificially raised queen from another source, the original colony now becomes two — and on the beekeepers', not the bees', schedule. This is not necessarily a bad thing — but it needs to be done in accord with the natural timing in the hive.

Taking The Honey in any Form or Quantity
Desired Without Destroying the Bees

In the days when bees were kept in *fixed comb hives* known as *skeps* or *log gums*, the only method of harvesting the honey was a *destructive harvest* — which involved destroying the entire colony in order to get at the honey-laden combs. The risk in harvesting honey so easily, however, is that the beekeeper may over harvest and leave the colony without sufficient resources to survive the winter season.

Limiting the Number of Drones

Drones consume honey but do not gather nectar, build wax, care for brood or contribute to the maintenance of the hive. Drones have always taken a bad rap from beekeepers, who maintain that aside from mating with queens, drones do none of the work involved in maintaining the hive, but cost much to maintain. And so, drones are said to have little or no value. From a honey producer's point of view, perhaps this is valid, but it ignores the potential effect that eliminating so much of the drone population has had on the genetic diversity of the bee population.

Controlling the Sex of the Bee

Beekeepers also found that controlling the size of the cells on the foundation could dictate the sex of the bee that was raised in that cell. This led beekeepers to manipulate the gender balance in the hive by providing foundation sized so that it promoted only the raising of worker bees. Since in some views the presence of drone bees is less than desirable, fixed-size foundation offered an easy way to have fewer of them.

Changing the Size of the Bee

At some point in the short history of modern beekeeping, people learned that a smaller cell would make smaller bees, and thus more bees per frame… and that a larger cell would make a larger bee, though fewer bees per frame. Perhaps the larger bee would be able to fly further, fly faster and be able to carry more nectar, resulting in the production of more honey? Foundation permitted reducing or enlarging cell size beyond what was found in natural combs. This is manipulation of a natural system at a deep level, and it appears to have contributed to the varroa mite's prolific success.

Chemical Contamination in the Hive—the New Worry

Since the advent of Colony Collapse Disorder (CCD) in late 2006, researchers have become increasingly concerned about the effects of the chemicals being used by beekeepers to eradicate pests in the beehive.

Many *acaracides* are extremely wax soluble, meaning that these compounds are absorbed into the beeswax comb of the hive —where they accumulate due to repeated applications over multiple seasons. This accumulation contributes to the mites' developing resistance to the miticides, while

696 **POPULAR MECHANICS**

EMPLOY ARTIFICIAL COMBS TO CONTROL SEX OF BEES

Beekeepers are manifesting interest in artificial combs made of aluminum, which are claimed to increase the production of workers and limit the number of drones. Another outstanding feature of the invention is that it relieves bees of the necessity of building wax combs, and thus enables them to devote all their energy to honey making. Being of metal construction, the combs will not melt in excessively hot weather. Furthermore, in the event of infection occurring in a brood, the cells may be thoroughly sterilized. In view of the request of the Department of Agriculture that keepers of bees market only extracted honey, so as to increase production by reusing the combs, the new aluminum devices seem particularly timely. They may be used indefinitely, are of standard size, and weigh about four ounces each. The cells consist of aluminum ribbons which are shaped by being passed between crimping rollers. By making the honeycombs double depth, they cannot be used for brooding. The brood combs, for the production of workers, are supplied with cells ⅜ in. in diameter. Since it is impossible for the queen bee to enlarge the holes, only females are produced in them. A special brood comb with cells ¼ in. in diameter is provided in which the queen may lay eggs for drones. This arrangement not only makes it possible to control the proportion of male and female bees, but also is said to permit blood control so that excessive inbreeding is prevented. Before being placed in hives, the combs are sprayed lightly with wax.

When it is considered that bees expend as much energy in making a pound of wax as they do in gathering 20 lb. of honey, the potential value of these artificial combs apparently should prove to be considerable.

Above Is an Aluminum Comb with Capped Cells Containing 15 or More Pounds of Honey. This Shows That Bees Accept the Artificial Devices

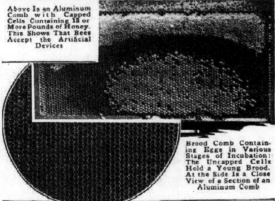

Brood Comb Containing Eggs in Various Stages of Incubation; The Uncapped Cells Hold a Young Brood. At the Side Is a Close View of a Section of an Aluminum Comb

the bees are sickened and stressed by the increasing levels of poison being absorbed into the comb, that special and essential structure where the bees store their food and raise their young.

Two common miticides are coumaphos and fluvalinate. Coumaphos, an organophosphate, is the active ingredient in CheckMite+, and fluvalinate, (tau-fluvalinate) a synthetic pyrethroid, is the active ingredient in Apistan.

Both of these compounds are *persistent*. Picture this chain of events with me:

The beekeeper treats the hive with CheckMite+—and treats it again next year and the year after. The mites by this time have developed a resistance to CheckMite, and so the beekeeper changes over to Apistan. The hive is treated this year and next year and the year after...and by this time the comb is quite old and dark and funky. The comb is then sent away to be recycled and made into brand new sheets of foundation. But these two miticides, along with other chemicals used by conventional beekeepers to treat their hives, survive this wax recycling process and they are now found in the wax of brand new sheets of foundation.

This means that a brand new hive, outfitted with apparently fresh, clean sheets of wax foundation is actually already contaminated with coumaphos and fluvalinate. The long-term, sublethal effects of these pesticides are still being studied—but initial research indicates that there are important effects on brood, drone virility and queen longevity. This contamination was brought to light during preparation for a study done by a research team at the University of Georgia to learn more about these long-term, sublethal effects. This study's focus was on four chemicals—fluvalinate, coumaphos, amitraz and copper naphthenate—that have been used to treat honeybee colonies in the USA.[2]

The research team at MAAREC (The Mid-Atlantic Apiculture Research and Entension Consortium) studied 887 wax, pollen, bee and associated hive samples, and found 121 different pesticides and metabolites contained within those samples, including coumaphos and fluvalinate.[3] They found 121 different pesticides contained in the hives, and especially in the wax.

In another research project at the University of Georgia, the College of Agricultural and Environmental Sciences studied the interactions between the most common varroacides — coumaphos and fluvalinate. This study showed that the toxicity of each of these chemicals was greatly increased when it was found in combination with the other. In some cases toxicity was seen to increase by 4 times, in other cases as much as 32 times.[4]

Cornell University's Pesticide Information Profiles (PIPs) provide detailed analysis of the biological and ecological effects of both coumaphos and fluvalinate.

A Few More Thoughts on the Use of Foundation

Wrong Shape

A less obvious concern, but still probably important to the bees, is the fact that a rectangle is not the shape of natural honeycomb. Bees build their comb from the top down — and in a rounded shape called a *catenary curve*. This is the same shape we would see if I handed you one end of a piece of chain and I held the other end and we looked at the curve, or drape, made by the hanging chain. This curve is both efficient and structurally sound.

You will see this shape in beehives where the bees have built their own comb. It is a result of *festooning* — the bees hanging in a gentle curve, seemingly holding hands — building their comb in that same shape.

Bees Don't Like It

One last and yet very important thing must be said about the use of foundation inside beehives: The bees don't like it.

Every beekeeper of any longevity eventually learns that given the chance, bees are apt to build natural comb in any open space left inside the hive by the beekeeper, even when foundation has been provided.

Here's a particularly brilliant illustration of the bees' distaste for foundation. A recent novice top bar beekeeper was worried that the bees he installed into his top bar hive would not want to stay inside the hive without a piece of foundation to work from. So he altered a top bar to hold a piece

This open-air hive shows the natural catenary curve made by bees.
Credit: Christy Hemenway.

of standard wax foundation, and then he installed a package of bees into the hive. He called several weeks later to confess what he had done, and to report that after building nine bars of beautiful natural wax, the bees still had not touched the foundation he had provided.

Often, beekeepers find that the natural wax their bees build when they get the opportunity to build without the constraints of foundation contains many drone-sized cells. This is then followed by the remark that the bees build too many drones. Why would bees build more drones than they needed? Is there such a thing as "too many drones?"

When bees are prevented from building drone cells by the fact that the foundation provides only worker-sized cells, it stands to reason, doesn't it, that they draw lots of drone comb when given the option to build natural comb? They are playing catch-up. There's no such thing as too many drones!

Does Standardization Make Sense?

When we view ourselves as outside nature — separate from it, different from it, superior to it — we act as if we are exempt from nature's laws. And this makes us callous and careless.

Two ideas are at play here: standardization and interchangeability.

Since standardized foundation began to be used consistently, many honeybee pests and problems have manifested to a frightening degree: tracheal mites, varroa mites, nosema and now the frightening disorder we call CCD. All these have gained a foothold, changing the face of beekeeping forever.

The dangers of standardization spring from disregarding nature's ways. Interchangeability allows us to support the nature's systems at work inside the hive:

- Being able to move bars from one hive to another to help a new beekeeper get started makes sense.
- Being able to *split* a hive — and move bars of natural beeswax comb from an existing hive into a new hive in order to preempt a swarm and increase the size of your apiary — makes sense too.

- Being able to offer a queenless hive a bar of open brood comb—the re-source they need to make a new queen—also makes sense.

While there are many things to commend the concept of interchangeability, wax foundation with its standard-sized cells, designed to raise only worker honeybees, is not one of those things. The most important natural system inside the beehive is the creation of natural beeswax comb—wax made by bees, for bees—and it needs to stay that way.

Nature does not hurry, yet everything is accomplished…

« Lao Tsu »

CHAPTER 3

Basic Bee Biology

Time and Temperature

Time and temperature have a great deal to do with honeybee biology. You can deduce a great deal of information about the status of your colony by knowing how long it takes for things to occur inside the beehive and what temperatures are required to enable the bees to accomplish certain tasks. You can then make sound decisions based on that information.

For instance — if you know how long it takes for baby bees to be born — and you know the date that your bees were hived, then you can predict approximately when you will begin to see an increase in the population of your hive. If you see eggs, then you can pinpoint that moment even more exactly because then you will have a more precise knowledge of when the queen began to lay eggs. Queens don't always start laying on day one — it can easily take a week, sometimes more, for a queen to be released from her queen cage and lay her first eggs.

If you install a package of bees and you notice drone bees during an inspection two weeks later, you will know that those drones were included in the package that you hived. Why do you know this? Because you know that it takes roughly 24 days for a drone to be born, so obviously those drones didn't hatch from your hive.

If you know that the egg stage of a bee's life cycle is three to four days, then when you see tiny, just-hatched larvae, you can do the math backwards and know roughly how long ago you had a laying queen. Knowing whether you have a healthy laying queen is very important, and yet you don't always see her in person when inspecting your hive. But you do see the signs of her activities and you can learn a lot from those if you understand the effects of time and temperature on the workings of the colony. This knowledge will keep you from digging through your hive over and over again determined to lay eyes on the queen when really there is no actual need to see her — if you can read the math of the hive.

But before we get any deeper into the math, first let's talk about the inhabitants of the hive…

Who's Buzzing in this Beehive?

Few societies, insect or otherwise, point out gender stereotypes so well as honeybee society. Since the female worker bees make up the vast majority of the population of the hive and do virtually all the work of the colony — cleaning, defending, foraging, feeding, caring for young — and the drones do none of these tasks, but spend their days idling away the afternoon in a *drone congregation area* (DCA) waiting for queens to come by to mate, well — you can see where that's going, can't you? Add to that the fact that the drone bee dies after "getting lucky" and mating with the queen, and you can see how beekeeping is a hobby just rife with bad puns and terrible gender jokes — always a source of great amusement!

But just who are the occupants of a thriving honeybee colony? The average honeybee colony is said to contain approximately 65,000 honeybees. This population consists of one queen, roughly 55,000 worker bees and, during the height of the summer season, approximately 10,000 drones (about 15% of the population of the hive).

Queen Bee

The queen is the only bee in a thriving, healthy hive that lays eggs. You can learn to recognize her by looking for several different features.

- Her thorax is very smooth, shiny and black.
- Her abdomen is considerably longer and more pointed than a worker bee's.

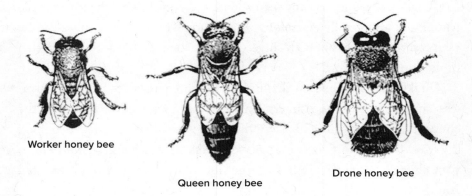

Worker honey bee

Queen honey bee

Drone honey bee

Worker, queen, drone—drawing.

Credit: EXtension.org. *Nest Occupants (Basic Bee Biology for Beekeepers)*. [online]. [cited September 3, 2012]. eXtension.org/pages/21745/nest-occupants-basic-bee-biology-for-beekeepers.

- Her abdomen is usually more of a solid color than striped like a worker bee's.
- A queen's wings are shorter—they will not reach the length of her abdomen.

She is frequently seen in one of two modes: standing fairly still on the comb and surrounded by a circle of worker bee attendants or, as is more likely while the beekeeper is inspecting the hive, you will just catch a glimpse of her moving rapidly across the comb, determinedly headed away from wherever you are looking.

Drone Bee

It's fairly easy for a new beekeeper to mistake a drone for the queen, as drones are also impressively large when compared to workers, but there are some differences that are easy to spot:

- A drone's abdomen is blunt and rounded, instead of slender and pointed.
- A drone's thorax is quite large as well, a drone having very well-developed wing muscles in the thorax. The capacity for swift flight is important to the drone's successful mating with the queen.
- Another identifying feature of a drone is the eyes. The eyes of a drone are so large that they cover almost the entire front of the head—so large that they actually touch in the center.

In the fall, the drones are not even suffered to come inside the hive for the winter, but are literally evicted by worker honeybees. Worker bees can be seen carrying drones out of the hive, dropping them on the ground and then prohibiting their return.

Unlike the other bees in the hive, drones are considered mostly harmless, since they have no stingers. This inability to sting contributes to the concept of drones' limited usefulness—as they cannot sting to defend the hive. But it also means that you can pick them up and hand them to your children safely—certain to garner lots of admiration from fascinated youngsters!

Worker, drone, queen—photo.

Credit: EXtension.org. *Nest Occupants (Basic Bee Biology for Beekeepers).* Photo © Zachary Huang. [online]. [cited September 3, 2012]. eXtension.org/pages /21745/nest-occupants-basic-bee -biology-for-beekeepers.

Worker Bee

Worker bees are usually the standard for comparison of the words "larger," "rounded" or "pointed." A worker bee can be recognized by these features:

- First, there are lots of them—85% of the hive will be worker bees.
- The thorax of a worker bee is a dull black, even slightly fuzzy. The younger the bee, the fuzzier it is—fuzzy yellow bees are only days old.
- Her abdomen and her wings will be approximately the same length.
- Her abdomen, in the case of most European honeybees, will be striped. Some are darker than others to the extent of being nearly solid, and some lighter to the same extreme, but as a rule, worker bees have stripes. And these stripes are not really colored stripes on the exoskeleton of a bee— but more of a revealing of the color of the bees' body from beneath a fine layer of hair, which as the bee ages, and the hairs wear away, creates a more defined stripe.
- A worker bee's eyes are small, round and separate from each other.

One of the problems with descriptions of this sort is that they are all comparisons. Beekeepers will tell you all day long that the queen is longer than the drone, and that a drone is bigger than a worker—but when talking about it in a classroom, it can be difficult to make anything of that information. Until you've opened a hive, held bars of bees in your hands and seen drones and workers side by side—or been able to compare the long, tapered abdomen of a queen to the shorter, striped abdomen of a worker or the blunt, rounded abdomen of a drone—this information doesn't really hit home.

Bee Math for Different Inhabitants of the Hive

- A queen bee spends 3.5 days in the egg stage.
- A queen spends 4.5 days as a larva, bringing the total to 8.
- This is followed by 8 days as a pupa in her sealed cell.
- A queen hatches on the 16th day.

- A worker bee begins life as an egg, and in 3.5 days she hatches.
- She then spends 5.5 days on the larval stage, taking us to day 9.
- Then she is capped—sealed in her cell—and spends 11 days in the pupa stage.
- She hatches on the 20th day.

- A drone bee begins its life as an egg, and he also hatches in 3.5 days.
- A drone exists in the larval stage for 6.5 days, taking us to day 10.
- A drone then spends 14 days in the capped pupa stage.
- He hatches on the 24th day.

Here is the same information, presented as a chart:

**The Length of Each Stage
of the Bee's Gestation Period in Days**

Caste	Egg	Larva	Pupa	Total Gestation
Queen	3.5	4.5	8	16
Worker	3.5	5.5	11	20
Drone	3.5	6.5	14	24

Credit: Christy Hemenway.

The time spent in each stage for each *caste* (type of bee) can vary by as much as a full day—in warmer temperatures things happen more quickly—a prime example of the important connection between time and temperature in beekeeping.

More about Queens

Oftentimes when you purchase the queen from a bee supplier, she is *marked*. That means that there is a tiny dot of paint, or a round, colored and possibly numbered disc applied to her thorax. If she has been marked in accordance with the International Queen Marking Code, the color of this paint dot indicates the year she was born.

International Queen Marking Color Code.

Color	For Year Ending In
white (or gray)	1 or 6
yellow	2 or 7
red	3 or 8
green	4 or 9
blue	5 or 0

Credit: Michael Andree. "Finding the Queen." Bee Informed Partnership website. [online]. [cited May 23, 2012]. beeinformed.org/2011/11/finding-the-queen/slide1-11/.

While this is helpful in keeping track of just how old your queen is, once you have a bit of experience with spotting queen bees you'll find that you won't need to put potentially toxic paint on your very important queen in order to be able to find her. I suggest that you simply develop the ability to spot a queen bee and eliminate the need to mark her.

Another technique used on queen bees is *clipping*. Clipping a queen is an attempt to thwart the swarming impulse of the hive. Tiny scissors are used to cut her wings so that she is unable to fly; then she can't leave the hive even when the colony is prepared to reproduce. However, the swarming impulse is how the colony, as a superorganism, reproduces—and the impulse is incredibly strong. So it is rare that efforts to stop swarming ever truly succeed. In a case where the original queen cannot fly because she was clipped, the colony makes its preparations to swarm in spite of this, and then swarms anyway, flying with one of the virgin queens that are raised as part of swarm preparation, and leaving behind the original mated queen.

There's an amusing story that relates to why it is so commonly believed that there is only ever one queen bee in a hive. The story goes like this: Have you ever wondered why, when you're searching for your car keys, they are always in the last place you look? Think about that for just a second... The answer? It's because when you find them, you stop looking!

The same thing frequently occurs during a hive inspection. The beekeeper does the inspection, looking and looking for the queen bee. Eventually she is found, and then what? The beekeeper stops looking for queens. So it's probable that 15 to 20% of hives have two queens, at least for a short time during the season. Contrary to the belief that all queens fight to kill other queens, it is possible for two queens, especially a mother and a daughter, to coexist side-by-side, both of them laying eggs and building up the colony.

The sexually active part of a queen bee's life is short. She mates only when she is very young — usually in the first two weeks of her life. She may take several mating flights, but after those flights have occurred and she begins to lay eggs, she never mates again. So it's very important that the mating process go off without a hitch during that early, prime mating period in the queen's life. Bees mate only in flight, not inside the hive. A queen bee receives and stores enough sperm in these early mating flights to fertilize all the eggs she will ever lay. She mates with many drones, and in fact the more promiscuous she is — in other words, the more drones she mates with — the stronger her pheromone is and the more loved she is by the hive. Ten to 20 drones is fairly common; the record I have heard is 40 matings!

MythBuster!

A commonly held myth is that there is only ever one queen bee in a hive. The photograph below shows that this is not necessarily the case.

Credit: A fellow beekeeper.

Long Live the Queen

Queens are essential for the survival of the hive. It's not that she is in charge — or directs any of the activity of the hive — the bees do that, guided by their amazing *hive mind* process. But because she is the only bee able to lay fertilized eggs that hatch into female bees, the ongoing success and survival of the hive depends entirely upon the health and fecundity of the queen. Queens are said to lay upwards of 1,500 eggs per day at the height of the brood laying season — this incredible rate of growth is crucial to creating the population build up that the hive needs so that they are able to gather enough forage in the summer to be prepared for the winter.

The presence of the queen's pheromone in the hive is what provides the colony with a sense that all is well — beekeepers call that being *queenright*. This pheromone is constantly being spread by the worker bees throughout the hive, and in the event that some accident or injury befalls the queen, the bees become aware of their queenless state very quickly. The level of queen pheromone in the hive dissipates quickly, literally within a few hours.

There are several scenarios where a new queen bee needs to be created. Some of these events happen intentionally (in the normal course of business), some of them are emergency replacement situations.

One of the most frequent causes of queen loss is the beekeeper. It is easy to accidentally roll a queen bee between combs, thus injuring her — crushing her beneath a top bar, or between bars — or worse yet, to have her fall off a comb while inspecting, land on the ground and then get stepped on. For this reason, I always recommend that you inspect cautiously and carefully, and always hold the combs over the hive while inspecting — a simple precaution that can help to prevent this last tragedy.

Swarming

When a colony is prepared to reproduce, they create swarm queen cells — usually several of them. From these cells will hatch new virgin queens, who will promptly strive to eliminate their competition, other virgin queens, by stinging them to death, often while they are still in their cells. The successful queen will fly to mate, return to the hive and then begin her life as the queen for the bees in the *remainder hive* — the bees that remained behind after a swarm, in the original hive.

Swarming is a reproduction event — this is how bees make more bees. This queen replacement process is completely intentional — and the existing queen in the colony participates in it by laying an egg in the vertical cell that was prepared especially for this purpose. To read more about swarms, see Chapters 5 and 6.

Supersedure

A medieval-sounding term that means to supplant, or take the place of the previous authority — this term conjures up visions of plots to overthrow the monarchy — a description that really isn't that far off. Should the bees become dissatisfied with their queen — whether she has become so old that she no longer lays enough eggs to keep the colony viable, or her pheromone levels are not strong enough to maintain a sense of cohesion in the hive, the sense of being queenright — they move to replace her.

The bees choose a freshly hatched larva and use that larva to make a replacement queen. They reshape the wax of its brood cell, turning it so that the cell is oriented vertically. This is the reason that supersedure cells are found on the face of the comb — since that is where young larvae will be found. That larva is then fed a steady diet of royal jelly, and she morphs into a queen bee.

Supersedure is a queen replacement event — but not a colony reproduction event. It is intentional on the part of the bees — but the queen does not participate, and in fact the bees will do away with her when they are certain of the success of her replacement. They may run her out of the hive, or *ball* her — a phenomenon where a large ball of bees cluster around her and create so much heat that they kill her.

Emergency Queen Replacement

Should the level of queen pheromone in the hive drop too low — due to injury, the purposeful removal of the queen or beekeeper error — the bees quickly become aware of this and begin an *emergency queen replacement* immediately.

It is possible for a hive that has suddenly lost its queen to replace her on their own — provided they have brood available in the hive at the appropriate phase of the larval stage. Again, a freshly hatched larva is best — and the bees use that larva, reshaping the wax cell and turning it so that the cell is oriented vertically, just like they do when superseding a queen. They feed that soon-to-be replacement queen the same royal jelly diet that a swarm or supersedure queen would be fed, and in 12 more days, they will have a new queen.

This is a life-or-death event for the hive. Without a queen, the hive cannot survive. If they are unable to replace their queen, the queen pheromone level in the hive will drop to where it no longer performs one of its major functions — the suppression of the ovaries of worker bees. Should this occur, the hive experiences a condition known as having a *laying worker*.

Laying Worker

In a thriving, healthy hive, a worker bee lays no eggs. Worker bees are considered female in that they have ovaries and they are able to lay eggs, but because they did not begin life as a queen, they have never mated and so they are only ever able to lay drone (male) eggs and never worker eggs.

Without worker bees, none of the work inside the hive will happen — no cleaning, no nursing, no feeding of babies, no foraging, no gathering of pollen or nectar, water or propolis resins. The number of drones soon begins to exceed the number of worker bees, and the whole hive is now in a downward spiral — a situation we sometimes call a *colony of lost boys*.

It is difficult for a hive to recover from having a laying worker. Beekeeper intervention is required. If the beekeeper has the resources available, he or she can provide the laying worker colony with bars of open brood comb. A bar of open brood provided once a week for two to three weeks can set this back to rights. The pheromones emitted by the open brood will begin to reset the balance in the hive, suppressing the egg production by the laying worker females. By the third week the pheromone balance will have returned to the point where the bees will become newly aware that they have no queen, and they will begin work to replace her, using the fresh larva provided in the comb the beekeeper has inserted in the hive.

A Hive Can Save a Hive—An Argument
in Favor of Interchangeable Parts

Should you find yourself with a queenless hive that has no very young brood in it — please understand this — that hive is helpless to produce a new queen via any of the natural methods described above. But you can help that hive to save itself — by providing a bar of comb containing open brood from another healthy, thriving hive.

This amazing ability of bees to create a new queen bee from a worker bee larva is the best reason I can think of for having interchangeable equipment — where the bars from one hive can be moved into any of your other hives — to resolve a life-or-death situation for your bees.

Temperature

Temperature affects the activities of a honeybee colony, both inside and outside the hive.

Some important inside-the-hive temperatures to know:

- The colony must be able to keep the brood at 93°F in order for it to hatch. Whether that means keeping it warm or keeping it cool depends on the time of year.
- The temperature of the cluster in a hive during winter, when no brood is being raised, is about 55°F.
- A colony must be able to heat the area where they are preparing to build wax to 91°F. This helps you decide how big a space to install bees in — based on the outside temperature.
- Typically bees don't fly below 48°F, though some breeds such as Russian bees, are said to fly in colder temperatures.
- You can hive warm bees in cold weather. However, you should not hive cold bees in cold weather. If a single bee's body temperature drops below 46°F the bee becomes sluggish or even paralyzed. This can create a serious problem when installing a new colony of bees into a top bar hive: If the bees are cold, the weather is cold and the top bar hive has no existing comb, the bees will be unable to reform into a cluster and the colony will die in a cold heap on the bottom of the hive. So if you must hive your bees on a cold day you need to start with warm (room temperature) bees.

Location and Climate

Two important factors outside the hive are also closely related: the geographical location of the hive and, by extension, the area's climate. Climate and weather vary significantly from region to region, even down to microclimates within regions. It is important when comparing beekeeping management practices to be cognizant of this. A beekeeper in the southeastern US faces a completely different set of circumstances than does a beekeeper in New England, or in the rainy Northwest. Significant differences are also found within regions, such as between coastal and inland or mountainous locations.

The climate of an area determines the *growing season*. What is in bloom, and when, determines where and when the bees *forage*. The symbiotic relationship between plants that need pollinating in order to set fruit and bees that need food in order to survive is a magical dance, integral to our entire food supply. It's one of the things you find yourself noticing more and more as a beekeeper — how everything is connected: plants, weather, bloom time, location, temperature, precipitation. It's as if, when you become a beekeeper, you also become one part meteorologist, one part botanist and one part entomologist.

For instance, here in Maine our growing season usually begins about mid-April. That's when we typically see the first dandelion; that's when the maples begin to bloom and the pussy willows begin to sprout their fuzzy catkins. In warmer climates, these events happen considerably earlier in the year — in some climates the growing season is virtually year-round. Bees in cold climates have a super short growing season in which to flourish, and yet there are even honeybees being raised in Alaska.

Different pests thrive in different climates and so different hive management techniques are inherent to some locations that are unheard of or little-used in others. This is why beekeepers have to be careful not to overgeneralize things such as beekeeping calendars — as they are extremely specific to the locale.

The Pollination Behavior of Honeybees

Pollination is the fancy name for the event that occurs when grains of *pollen* from the flowers of one plant are moved to the reproductive organs of another plant of the same species. You could describe it as the plants having sex. Pollination almost seems accidental, after all it's just bees bumping up against flower parts, and yet it's crucial to the way the plant world works.

Honeybees are excellent pollinators because when they exit the hive, they concentrate solely on one type of plant on that foraging excursion. In other words, a honeybee that leaves the hive and visits a dandelion will next visit a dandelion, and then another and another. A honeybee that goes out to forage and visits an apple tree will then proceed to visit nothing but apple trees on that foraging trip away from the hive. A honeybee does not travel from dandelion to apple — and you don't see dande-apples!

The Stinging of Bees

As a rule, honeybees are not aggressive. It is only the female worker honeybee that is even able to sting. Drones, the male bees, have no stingers. And a queen bee, while she does have a stinger, only uses her stinger to eliminate another queen.

Stinging is an expensive defense mechanism for the bees. A honeybee can sting only one time in her life, because when she stings, she dies. Stinging is a kamikaze mission on her part — so she does not sting frivolously.

The reason that a honeybee can only sting once in her lifetime is this: A honeybee has a *barbed stinger*. And when that stinger penetrates your skin, your bee jacket or anything else, the barb causes the stinger to lodge there. When the

The aftermath of a honeybee sting—for the bee.
Credit: Copyrighted photo by Kathy Keatley Garvey.

bee is then brushed off or tries to fly away, her venom sac is pulled out of her abdomen along with the stinger, effectively disemboweling and certainly killing the bee.

So stinging is not something that a honeybee does for fun. Knowing this makes it a little easier to understand bee behavior. If you give them good reason, honeybees will sting, but as a rule, it is more in their self-interest not to sting.

So, why does a honeybee sting? The primary reason for a bee to sting is in the course of protecting the colony's two most precious resources — their brood and their food supply. The incidence of stinging, and bees' irritability, may increase in stressful situations such as during a *nectar dearth*. Hives that have been under attack by skunks or bears are also likely to be defensive. And some bees are just more irritable and less docile than others. On the highly irritable end of the scale you will find Africanized honeybees, and on the docile end, the Italian honeybee.

If you put your hand down on a worker bee, step on her or crush her between a top bar and your fingers, she will also do her best to sting you then — even as she dies.

Dark Colors

I don't claim to know the real reason why bees seem more likely to sting dark colors. I don't know that I hold with the anecdote you often hear that they think you are bear, and anyway it doesn't hold true in every single case. But if you'd like to increase the odds of not getting stung, it seems to make sense to wear light colors.

Strong Scents

It also makes sense not to wear heavily scented body care products or cologne — as that has been known to cause an aggravated reaction. So has the smell of bananas on your fingers, as the bees' *alarm pheromone* smells very much like bananas.

Not All Bugs that Fly and Sting are Honeybees—or Even Bees

Yellow jackets, wasps and hornets are not honeybees. Yes they all fly, and yes they all can sting — but yellow jackets and hornets and wasps seem to look at life with a completely different attitude than honeybees. Wasps, hornets and the like can be aggressive seemingly just for the heck of it — and there's a crucial biological, physical difference between the two…

Yellow jackets, wasps and hornets have *smooth stingers* so they are able to sting repeatedly without dying from it.

As a thinking beekeeper, you are likely to find yourself providing a public service by educating people about honeybees, their biology and their "sting policy." Every little bit helps — and the more people that understand that bees are beneficial insects, the more support we can garner for their protection.

Are You Really Allergic?

Reacting to a bee sting does not mean you are allergic. A bee sting contains bee venom. Venom is designed to cause a reaction in the victim. If you are stung by a honeybee, you should expect your body to have a reaction — a local reaction — to the venom contained in the sting.

Local

Likely local responses include pain from the event of the sting itself, redness, swelling and itching. The swelling and itching may be fairly intense, and they may last several days. These are normal responses to a bee sting.

A warning regarding swelling is appropriate here: If you are stung on the hand or arm and you are wearing a ring — remove it immediately! If your fingers begin to swell while you are wearing a ring, that can be uncomfortable at the very least and quite dangerous at the worst.

The amount of redness, swelling and itching varies according to the individual who has been stung, as well as the location of the sting and the amount of venom the bee was able to pump from her venom sac into the victim before the venom sac was removed or before the muscles pushing the venom into the sting site ceased to contract.

Systemic

Sometimes people worry that because they swell after a bee sting they are having a life-threatening, allergic reaction to the sting. This is unlikely. A true allergic reaction involves *anaphylaxis* — a severe, whole-body response to a substance that has become an allergen. This is a life-threatening situation and should be taken very seriously — but it is very different from a local reaction.

> ## A Frustration for Beekeepers
>
> This ability to sting multiple times, combined with the more aggressive attitude of these "Not-Honeybees," can make for some very bad public relations situations for honeybees.
>
> The media's inability to discern the difference between honeybees and other stinging insects means that any and all flying, stinging insects are portrayed in news reports as frightening and dangerous. Combine that with the delight that Hollywood takes in the horror genre with movies called *The Swarm* and *Killer Bees* — and it's no wonder that many people are afraid of bees.

The symptoms of an anaphylactic reaction happen quickly—within minutes or even within seconds. They can include difficulty in breathing, difficulty in swallowing, anxiety, confusion, dizziness and light-headedness, hives, extreme itchiness, redness, slurring of speech, wheezing, as well as abdominal pain, nausea, vomiting or diarrhea, among other things.

The proper response when someone has a severe allergic reaction to a bee sting is to get immediate emergency medical help. If the person has a known allergy to bee stings and has medication or an EpiPen, help her or him to take or use it.

An EpiPen in its protective case.

Credit: Photo by Sean William, licensed under the Creative Commons Attribution-Share Alike 3.0 unported license. [online]. [cited September 10, 2012]. en.wikipedia.org/wiki /File:Epipen.jpg.

Unpredictable

How a person responds to a bee sting can be unpredictable. Many beekeepers, over time, develop immunity or a resistance to the effects of bee venom. Some will tell you that the first sting of the season is different from a sting received later in summer or in the fall. Some beekeepers accumulate hundreds of stings in a season, some get stung once or twice, if at all. It's different for everyone.

A person's reaction can also change over time. There are anecdotes of people who kept bees for years without any mishap at all, then suddenly developed a more serious, even anaphylactic reaction to a bee sting. It is because of that potential for change that I include this warning: When you plan to work your bees, even if you work the hives alone, it makes good sense to let someone know that you are going out to get into the bees. Chances are very good that no ill will befall you—but on the off chance that something does, it would be brilliant if someone knew where to look for you!

Dried Bee Venom

Another potential way for someone to develop a reaction to bee venom has nothing to do with actually being stung by a bee. If a person is exposed to bee venom, but not enough of it to develop immunity to it, there is a chance of developing a very serious reaction.

This can occur if a person is exposed to flakes of dried bee venom, which can remain on a beekeeper's bee suit and then get into the air in the person's environment. This doesn't happen often—but it's worth noting. The best way to avoid this reaction is to wash your bee gear fairly regularly.

Another option is to get everyone in your household interested in bees and get them all stung regularly enough that they develop some immunity!

When you go in search of honey you must expect to be stung by bees.

« JOSEPH JOUBERT »

II

when to do what—
and why

CHAPTER 4

Your Top Bar Hive

While top bar hives are by no means a new technology, they are a significant departure from the conventional square-box hive which has dominated the US beekeeping landscape since the Civil War. So there could be said to be two camps, if you will. One camp carries a belief in a long-standing tradition, and the other a desire to do something different because what we've been doing for the last 150 years doesn't seem to be sustainable.

This has led to a swirling morass of information marked by strong feelings and high passion. Emotions confuse the matter for the novice, who has no basis for comparison and often finds it difficult to understand why people are so adamant about their own personal choices in beekeeping. In this chapter I will do my best to explain top bar hives clearly so that you can compare and choose the beekeeping method that makes the most sense for you.

Top bar hives have been made of all sorts of materials and containers — from baskets and flowerpots to recycled scrap wood, slabs of trees cut from logs, thin branches lashed together, even plastic barrels cut in half. A top bar hive can be elegant in its very simplicity. It takes advantage of the bees' natural cavity-nesting behavior; it requires little else but a space of adequate volume to suit the bees and bars of some type from which they are willing and able to hang their comb. The bars, by being removable and inspectable, make a top bar hive a movable comb hive, desirable since it allows the beekeeper to check on the progress of the hive, permits the diagnosis of pests and diseases and allows for the removal of honey without destroying the balance of the colony. In some places in the world where beekeeping is regulated, movable combs are also a legal requirement.

Of course, some containers and some top bars are more effective than others. Lots of folks have experimented with the shapes and sizes of spaces that bees will willingly occupy.[1] Some tendencies and preferences have been identified, but not to the extent that the bees' behavior can be predicted with any consistent accuracy—in fact, the many wacky places where bees have been found living are the stuff of many a bee story—not unlike a "fish story."

All this is to say top bar hives need not be precision-built pieces of woodworking—after all, beekeeping is not rocket science. There are plans available to download from many sources on the Internet; there are plans in books; there are hives built from scratch and from imagination by creative beekeepers; and there are currently hives being commercially produced by several small businesses, including my own, Gold Star Honeybees®.

The Important Elements of a Top Bar Hive

The three crucial elements of all top bar hives are the *cavity* of the hive body, the *top bars* themselves and the type and location of the *entrance*.

The Cavity

The volume of the cavity that the bees will occupy has to be large enough to house a colony of an appropriate size for the climate of the locale. It must allow for a large enough brood nest and adequate food stores for them to survive the region's winter conditions. A top bar hive containing 30 bars seems to work well for most climate scenarios in North America.

The Top Bars

The top bars rest across the top of the hive cavity, and from them the bees will draw their comb. The top bars should be designed in such a way that they can be removed from the hive body easily, examined and returned to the hive without crushing bees.

**A Gold Star Honeybees®
top bar.**
Credit: Christy Hemenway.

The most important requirement of a top bar is that it must offer a good comb guide for the bees. I've examined and experimented with many different ways of making top bars, and how the bees have built on them, and I've concluded that the most effective comb guide is a beveled point.

Extending the length of the guide for the entire working portion of the top bar also helps to prevent the bees building comb that curves sharply off the bar when they reach the end of a shorter guide.

Granted that the bees may choose to completely ignore all our best laid plans in any case, but a beveled point that runs the entire length of the working area of the top bar has worked best the majority of the time.

A comb guide of this type can also be an integral part of the hive design and serve to center the bars — preventing them from sliding back and forth over the edge of the hive. This helps to keep them from getting in the way of the roof when it is being put back on. Be aware though, that the guide portion of the top bar, the pointed edge of the bevel, where the bees will attach their comb, should angle in from the hive body, so that it does not touch the

Spacers—skinny-wise and flat-wise.
Credit: Christy Hemenway.

side of the hive, as the bees need that all-important ⅜ of an inch of bee space at each end, so that they can walk between the sides of the comb and the inside of the hive body.

The width of the bars inside a top bar hive have much to do with the success or failure of the bees to build their combs in such a way that it can be removed for inspection. In general, the width needed for brood comb is between 1¼ to 1⅜ inches. But bees will build honeycomb up to two inches wide.

This leaves the top bar beekeeper with a dilemma. In the interest of allowing the bees to build natural comb, it would seem prudent to create different widths of bars, but this can be frustrating since it's difficult to predict how many bars of what size a colony would utilize, and where and when they should be placed in the hive.

A compromise is reached by making all of the bars suitable for brood comb — 1⅜ inch — and then utilizing thin wooden strips called *spacers*, designed to be placed between the bars in order to move the bars apart to ac-

The top bar on the right squashes fewer bees.
Credit: Christy Hemenway.

commodate the bees building wider honeycomb. Spacers measuring ⅛ inch by ½ inch offer two options — inserting them into the hive skinny-wise adds an additional ⅛ inch of space, inserting them flat-wise adds ½ inch — essentially creating three different widths of top bar, which solves this problem reasonably well.

A bee-saving design feature concerns the vertical height or thickness of the bars — i.e., the vertical sides of each bar, which touch the bars next to it. This dimension should be fairly small; ⅜ of an inch is plenty. The thicker the bar from top to bottom, the more difficult it is to keep from squashing bees when returning the bars to the hive during an inspection.

Another design feature that my experience has caused me to appreciate is that the ends of the bars should rest across the top of the hive body's sides and extend beyond the sides a bit. Most especially, they should not be set down inside any kind of surrounding lip or frame.

This "across the top" bar placement allows the beekeeper to separate each bar with a quick twist of the hive tool, and then to pick up each bar

Top bars extending over the edge—photo taken from below to show extension.
Credit: Christy Hemenway.

Center side entrance top bar hive.
Credit: Christy Hemenway.

easily by its ends without having to dig down inside a frame to pry the bars up and out.

The Entrance

Generally speaking, there are two different styles of top bar hives — those with a *center side entrance* and those with an *end entrance*. The biggest consideration when choosing between the two is probably the beekeeper's personal preference, but the two styles require different management techniques mentioned below.

An advantage of having the entrance in the center of the long side of the hive is that the colony can be installed on eight to ten bars in the center of the hive, between two *follower boards*. This offers some protection from the elements on each end due to the air space created between the follower boards and the hive ends. This is beneficial during a cold spring, when weather is often unpredictable.

The size of the entrance to the hive should be adjustable by some method. It must be able to be made large enough to accommodate the for-

End entrance top bar hive.
Credit: Christy Hemenway.

aging activities of the busy colony at the height of the season. It needs to be able to be made small enough to be defended easily when the colony is young and small. In a center side entrance hive, round holes, which can be plugged by corks when not in use, work well. A slot that can be blocked by a square piece of wood with a small hole in it works well with an end entrance hive. The entrance should also accommodate the installation of a mouse guard.

It is my experience that the bees don't seem to care much about placement of entrances, but entrance placement definitely affects the beekeeper's management of the hive, and particularly the use and placement of follower boards.

Management Differences due to Entrance Location

Bees in a center side entrance hive will build their colony in one direction, like those in an end entrance hive — beginning against one follower board and continuing toward the opposite end of the hive. But when they've done this, it leaves one third of the hive cavity empty — the one third on the other side of the follower board next to where the brood nest was started, the space where the feeder was originally installed.

To encourage the bees to continue building in the same direction and storing their honey at one end of the hive, the beekeeper does a *mid-season shift*: the empty bars above the feeder are removed; the entire hive is shifted into that space; and the empty bars are moved to the opposite side, so that the bees can continue to build their honey stores in that direction. The follower board with the feeder access hole is also moved, and the feeder as well (see Chapter 6).

It's important to manage any top bar hive so that the honey is stored only on one end of the brood nest. This is a winter management concern. If the honey stores are located on both sides of the brood nest, then, during the winter when the bees have consumed all the honey in one direction, they may be unable to cross the empty space to get to honey on the other side of the hive. This can cause a colony in a cold climate to starve.

Follower Boards

An end entrance hive will usually have only one follower board, which is set eight to ten bars back from the entrance when starting a colony, giving the bees eight to ten bars of space to build in, and then moving toward the back of the hive one or two bars at a time as the colony expands.

Center side entrance hives typically have two follower boards which act as false ends of the hive and follow the size of the colony. When the bees are installed, the follower boards are set eight to ten bars apart, centered over the entrances, again giving the bees eight to ten bars of space to begin building in, and expanding the hive one or two bars at a time.

Roof

A top bar hive needs a roof that protects the top of the top bars from the elements. This has been accomplished by many diverse methods. I've come to appreciate both the look and the practicality of a gable roof, because it sheds rain and snow and creates an attic space above the top bars. This space is useful for storing spare top bars and spacers, and even your hive tool. In the

The follower boards "follow" or contain the colony.
Credit: Jim Fowler.

winter the space can be filled with insulating material, and in the summer, a gable roof allows for air movement above the bars, cooling the hive.

Landing Boards

End entrance hives typically have a landing board, while center side entrance hives typically do not. Usually the first comment made in any discussion about landing boards is that trees don't have landing boards.

I don't know that the bees care one way or the other, but my observations have been that a landing board's biggest effect is on the appearance of the activity at the entrance. At a hive with a landing board, it is fun to watch the bees come in for a landing and walk in.

But it is also fun, in the case of a center side entrance hive with no landing board and where the entrances are drilled directly through the side of the hive body, to see the bees hover gently outside the doors and then suddenly zoom straight into the hole. They are also quite capable of landing on

An observation window allows a low-stress peek into the hive.
Credit: Christy Hemenway.

the side of the hive body (think: tree!) and then walking into the entrance that way.

So while some beekeepers have strong feelings about the need for a landing board, I think this is another matter where the beekeeper's personal preference holds sway.

Observation Windows

While this is not a requirement, it is simple to install an observation window in a top bar hive. A window is educational, fun to have and so helpful in making preliminary checks of the hive that it almost seems silly not to have one. Be certain that there is also a shutter to cover the window when you are not peeking through, as the bees prefer dark cavities.

Other Significant Features of Top Bar Hives

The most significant and sometimes controversial difference between top bar hives and conventional, square-box hives concerns the use of wax foundation. A top bar hive is a foundationless hive. It does not require or support the use of wax foundation—in fact, the typical thinking beekeeper abhors the idea of wax foundation, as it thwarts the bees' ability to craft their nest according to their own instincts where cell size is concerned—

specifically as pertains to the bees' ability to build worker-sized cells and drone-sized cells.[2] Additionally, since the advent of the varroa mite and the heavy chemical treatments used by conventional beekeepers in an attempt to control them, the foundation commercially available today has been contaminated (due to recycling of chemically treated combs) with miticides and other toxins that go against the green and natural focus of most top bar beekeepers.

One further note on the bees' building of comb: The existence of a midrib, created by sheets of foundation inserted vertically in frames, causes bees to have to build out from the center. They do it, but it's not the way they would do it naturally, as in nature there would be no center point from which to start.

Another controversial difference between top bar hives and conventional Langstroth hives, at least in the US, concerns the orientation of the hive, i.e. horizontal versus vertical. Bees are cavity nesters and in nature are well able to utilize whatever cavity they occupy, regardless of its orientation, starting their comb at the top of the space and building down.

It has been said, quite emphatically by some folks, that "bees must move up." But in fact, bees occupying a cavity in nature attach their comb to the ceiling of the cavity they occupy and then build down—lengthening the comb by hanging in a catenary curve shape to fill the space available. So while they may be found in a vertical cavity, they build their natural comb down from above. But the Langstroth hive and others like it, are made up of a series of stackable boxes, with new boxes being *supered*, or added to the top. This causes the colony to have to move vertically inside the hive in order to build additional combs.

The movement of air and moisture inside the hive differs between conventional hives and top bar hives. In a conventional hive, the frames are purposely spaced apart because the bees must be able to move vertically between the boxes. This also affects the movement of air inside the conventional hive. Air enters the hive at the bottom entrance at the front and travels upward, where it can be vented by leaving an entrance open at the top. In the winter, this has a specific effect on the movement of moisture as well. Moisture travels upward to the top of the hive and condenses at the top of the hive. In winter it freezes there. It is common practice for the conventional beekeeper to insert an insulating material of some sort above the inner cover of the hive to absorb this moisture. The danger lies in having moisture drip down on top of the brood nest in spring—this moisture can chill and kill the brood. It is often said that the biggest dangers to beehives in winter are not from the temperature, but from moisture.

But in a top bar hive, the top bars all touch each other—forming a solid surface above the bars, similar to the top of a cavity found in the interior of a tree—with all the bees below. No air moves upward between the top bars; air circulates horizontally throughout the hive. This is a major difference!

The type of moisture problems experienced in a conventional Langstroth hive do not occur in a top bar hive as there is no opportunity for moisture to condense above the colony or over the brood. Moisture may sometimes condense on an observation window if there is one—but as bees need water inside the hive, this does not present a problem, and in fact may be helpful.[3]

The Gold Star Top Bar Hive

The length of the top bar, the volume of the hive cavity and the appropriate ratio between the weight of a full bar of honey and the amount of attachment of the comb to the top bar—all these concepts were taken into consideration in the design of the Gold Star top bar hive (shown on the next page).

On Interchangeability among Top Bar Hives

In the interests of thinking beekeepers being able to support each other, and to help to proliferate this sustainable, natural, chemical-free method of beekeeping, it behooves us to improve our ability to interchange equipment—especially the top bars—between hives.

Interchangeability

- matters when we need "a hive to save a hive;" we move bars of open brood from a queenright hive into a hive that is queenless or has a laying worker problem. See Chapter 3 for deeper detail on this topic.

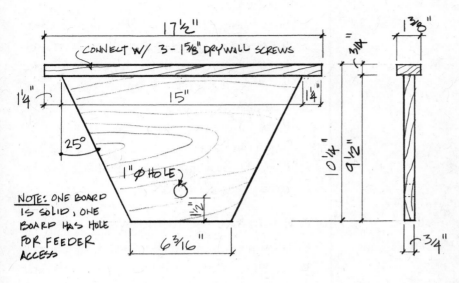

17½"

CONNECT W/ 3 - 1⅝" DRYWALL SCREWS

1¼"

15"

1¼"

25°

1" ∅ HOLE

NOTE: ONE BOARD IS SOLID, ONE BOARD HAS HOLE FOR FEEDER ACCESS

6³⁄₁₆"

½"

10¼"

9½"

1⅜"

¾"

FRONT VIEW   END VIEW

↑ FOLLOWER BOARD (BUILD TWO)

NOTE: ALL HIVE PARTS PRE-DRILLED FOR DRYWALL SCREWS ARE TO BE COUNTERSUNK

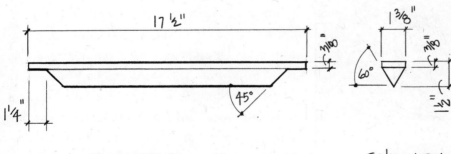

17½"

1¼"

45°

3/80"

SIDE VIEW   END VIEW

1⅜"

60°

3/80"

½"

↑ TOP BARS (BUILD 30)

Gold Star top bar hive dimensions.
Credit: Gold Star Honeybees®.

Warre hive.
Credit: David Heaf.

- matters when we need to share honey to feed bees — honey from top bar beekeepers whose treatment-free mindset and methods are similar to yours, so that the honey is safe for your bees. See Chapter 5 for more detail about feeding honey to your bees.
- helps more green beekeepers get started by populating a new hive with a *nuc* (starter colony) of bees from an existing top bar hive, instead of using package bees.

A Brief Note regarding Warre Hives

Before moving on, I'd like to quickly address Warre Hives. Developed by Abbé Émile Warré (1867–1951) and sometimes known as "The People's Hive," a Warre hive is often described as a vertical top bar hive.

Warre hives do indeed have top bars (bars that rest across the top of a cavity), and they do allow the bees to make their own natural beeswax — the most important feature of any beehive, in my opinion. However, the practice of lifting the entire hive in order to *nadir* or add a box to the bottom of a vertical stack strikes me as impractical if not impossible. I've seen some ingenious ways of accomplishing this feat, but they mostly seem to require some fairly elaborate equipment that represents a significant departure from the low-tech simplicity at the heart of top bar beekeeping.

Two other features make a Warre hive very different from a top bar hive:

1. The top bars in a Warre hive do not touch but are spaced a little less than ½ inch apart. This

causes a significant difference in air movement and in management style, and mimics the Langstroth hive more than a top bar hive.

2. The bars are often nailed or otherwise anchored into place, making the hive a fixed comb hive. This removes the beekeeper's ability to inspect the comb, which to me seems imprudent in the interests of following the progress and health of the colony, not to mention that fixed comb hives are not considered legal in many places in the US.

I appreciate that both Warre hives and top bar hives fall into the alternative hive category and that both are considered *foundationless hives*—and this makes them similar. However, I've got no personal experience with keeping bees in a Warre hive and since they are also significantly different in important ways, I would encourage interested readers to seek in-depth information about Warre Hives from a source more familiar with their specific management.[4]

CHAPTER 5

On Getting Started
with Your Own Top Bar Hive

The logical first step, (after reading this book, of course!) toward becoming a top bar beekeeper is to obtain a hive. In Chapter 4 I discussed different styles of top bar hives and the features of each. Here I will talk about ways of obtaining the hive of your choice.

Doing It Yourself

The number of plans for building a top bar hive that you can download from the Internet today (2012) is nothing short of amazing. Many of them are quite effective — but some of them aren't much more than ideas sketched on the back of an envelope. Look for testimonials speaking to the effectiveness of a hive built from such free downloads.

I may be a little more obsessive than some about using non-toxic products in a hive, but there are some things I think it's very important to avoid using in the construction of a top bar hive. They include

- plexiglas — it warps and off-gasses
- laminated wood products — the glues off-gas and most contain formaldehyde
- silicon caulk — it is toxic and off-gasses

You won't find any of the above in Gold Star Honeybees'® equipment.

Also remember the old adage "you get what you pay for." It's great and it's smart to save money — but if you build a home for bees, put them in it and then discover a problem with the design of the hive, it can be very difficult to change the box once it contains bees. If the flaw is serious enough, it can cost you the bees and a full year of beekeeping. Some top bar hive suppliers have years of experience with their product in the field and have refined their design to a high level. (Please see the design diagram on page 59 in Chapter 4.) If the do-it-yourself route is the one you prefer to travel, check into what's available from an established company in the way of DIY kits. You will not regret the extra leg up an experienced supplier can provide with a well-thought-out kit.

Each cycle of discovering and then implementing a hive design improvement takes a full beekeeping season — an entire year — so it's great to be able to take advantage of others' mistakes and learn from them. This learning speeds things up for you considerably, and there's little sense in reinventing the wheel when honeybees need help and protection.

Buying a Ready-Made Top Bar Hive

As of this writing, there are at least half a dozen small businesses offering top bar hives for sale. Their prices, quality, effectiveness of the hive and customer support services all vary. Carefully consider all of these things when making your decision because here again, once you've committed to a supplier, is isn't easy to make changes once the bees are in the hive.

If the hive you're considering sounds too good to be true, you can be pretty certain that it is. If the price seems unbelievably cheap, there's a reason. Look for an established company who is in business for the long term — someone who will be there next year to answer your second- and third-year beekeeping questions and one who is willing to talk with you and help you learn when issues come up, whether with the bees or with the equipment.

Where Do the Bees Come From?

A hive can be can populated in a number of ways. One way would be to collect a swarm. Another way is to purchase a package of bees. Or you could *split* a thriving top bar hive—your own or another beekeeper's (provided your equipment is interchangeable).

Then there are some more extreme methods, such as cutting up a *nuc* (nucleus colony) designed for a Langstroth hive—or by relocating bees from inside the walls or roofline of a building.

A Swarm

In an ideal natural world, hives would all be started from a large, healthy swarm. Since swarming is the bees' natural method of reproducing their

Capturing a swarm of honeybees.
Credit: Christy Hemenway.

colony, it has all the strength and vigor that nature can provide. A swarm is a complete, turnkey, start-up colony. It consists of approximately half of an original colony that has left home in search of a new beginning. Swarming bees are well organized and seriously motivated, with their *hive mind* consensus, to find a new cavity and begin building honeycomb as a place for their queen to begin laying. The bees in a swarm are all the right ages to do the tasks needed in their new home. A swarm can make wax and build comb at an amazing pace — some have been known to make more than one full bar per day when they are getting started.

Capturing a swarm is not always an easy task and never predictable, but it can be an inexpensive way to populate a top bar hive. One thing to consider is that the bees may have swarmed from a hive that had been treated with chemicals — but that is tempered somewhat by the fact that they were strong enough, and vigorous enough that they swarmed.

However, swarms are an event of nature, and you may or may not be able to connect with one.

A Package

Another method of obtaining bees, less natural but still effective, is to purchase a package of bees. Package bees can be ordered, along with a queen bee, from a honeybee supplier. You can locate package bee suppliers on the Internet, through your local beekeeping association or through your personal beekeeping mentor.[1] One thing to be aware of when ordering package bees is that, as of this writing, many of the apiaries supplying package bees in the US do use multiple chemical treatments. You can ask them about that.

In addition, some of the hives used as source hives to make packages have been on the pollination circuit earlier that year, so they may be very worn, stressed bees. You can ask them about that, too.

Don't be shy about asking these questions when purchasing package bees; the paradigm is definitely shifting toward sustainable, chemical-free beekeeping. As thinking beekeepers, we can best support and promote that trend by patronizing treatment-free apiaries as this trend grows.

When ordering packaged bees, plan well in advance. January is not too early to be shopping for bees! Most suppliers will be sold out by March or by April at the latest. The delivery date of your bees will coincide with the growing season in your area, which is dictated by the local climate. For example, in places where it snows in winter and the growing season begins in late April or early May, these months will be the likely arrival time. Conversely, if the growing season in your area begins much earlier, bee arrival will be earlier as well.

The package may have traveled some distance to get to you — perhaps a couple of hours, perhaps a couple of days. So it's important to pick the package up from your local supplier or your local post office as soon as it arrives and to have your top bar hive ready and waiting to install them as soon as is practical. The bees will be anxious to fly and to start foraging.

A package of honeybees.
Credit: Christy Hemenway.

Treat the bees like you would a new puppy: Put the package in a safe place, out of the sun — not too hot, not too cold — with good ventilation. Don't open the package until you have thought through the installation procedure described later in this chapter, gathered all your tools and are ready to follow all of the steps.

What's Inside a Package?

Bees: Three pounds of bees — about 10,000 honeybees — will be contained in the screen box. They will be clustered, i.e. gathered together into a mass or ball, around the *queen cage* and the *feed can*. There will likely be some bees dead on the bottom of the package; don't be overly alarmed by this. Bees only live about 42 days, so obviously some bees may die in transit. It's considered an industry standard that up to one inch of dead bees on the bottom of the box still leaves the necessary number of bees in the package to launch a thriving colony. Any more than one inch, however, and the backyard beekeeper should contact the supplier about their replacement policy. When you go to pick them up, you may find you have some *hitchhikers* as well — bees that are clinging to the outside of the package. Be careful not to squash one of these hitchhikers when you are handling the package, since they will do their best to sting you as they die.

One Queen in a Queen Cage: The queen will be confined in a separate *queen cage* inside the package. The queen cage is a small wooden box with two or three circular spaces in it. One of these circles will contain a thick white sugar *candy plug*. This candy will be blocking the entrance through which your queen will eventually exit the cage. The other two circles will contain the queen bee.

Don't be surprised if you find more than one bee in your queen cage. It is a common practice for there to also be several *attendant bees* included in the queen cage — these bees help to groom, feed and clean the queen bee, who performs none of these tasks for herself.

The queen cage offers some important safety features. The first is to contain the queen so that you know where she is and keep her safe until she is safely installed in the hive.

The queen in her cage, with attendants.
Credit: Christy Hemenway.

The second safety feature is necessitated by the fact that the queen that comes in a package is not the packaged bees' natural queen. The *queen pheromone* emitted by the queen bee helps to keep the colony together as a unit, but until the bees have come to accept the pheromone of this new queen, there is a chance that they may try to kill her. Because of this, another function of the queen cage is to allow for a slow release of the queen into the colony, giving the bees several days to become accustomed to her. The time it takes for the bees to eat their way through the sugar candy plug blocking the exit, releasing her into the hive, allows enough time for the bees to accept her.

Can of Feed (Sugar Syrup): During transport, the bees in the package are provided with a source of liquid food, contained inside a metal can that is held securely in place by a frame built into the package and the package lid. There will be a couple of tiny holes punched in the bottom of this can through which the bees can feed. These holes are quite tiny and release only a very little bit of syrup. The holes in the feed can are small of necessity, since the package is exposed to a lot of jostling and vibration during travel and would otherwise leak, making a mess and making the bees sticky, not to mention leaving them without food.

A Split

Another method of populating your hive is through a *split*. A split is just what it sounds like — one hive split into two. With a split the hive starts out with more than just loose bees; the bees will also have resources — drawn wax comb containing brood and food. A split is a small *nucleus* or starter hive. Splitting gives the bees a strong start, and their hive can really take off.

If you have a hive already and can split it into two — and have a second hive ready to accept it, you can increase your apiary from your own successful stock in this way. If you are a beginning beekeeper and have no existing colony but can find a willing beekeeper with interchangeable equipment — and bees that are ready to be split — then you may be able to obtain one of these and be well on your way.

Hack'n'Slash of a Nuc

Langstroth nuc colonies are an extreme way of populating a top bar hive. They have few advantages as the Langstroth colony is likely to have been treated with chemicals, and in any case, the comb will probably have been drawn from contaminated foundation (see Chapter 2). A treatment-free Langstroth nuc, raised on natural wax or *small-cell foundation* wax, would be a huge improvement, but it is still not in your or the bees' best interest to use a Langstroth nuc, since it requires that you literally take the frames apart, and then cut the comb, which will contain eggs and larvae, to fit the hive body of your top bar hive. This is very hard on the bees, and can be very stressful for you as well! The exception to this would be if your top bar hive is made the same size as a Langstroth nuc — then the hive parts would be interchangeable. But if that's the case, I'd ask why not just keep Langstroth hives without using foundation?

Relocating Bees from Buildings

This is another extreme way of populating your top bar hive. I won't spend much time on this method of obtaining bees — since it usually necessitates opening the building and the potential liability involved is high. Doing this very thing — which I called "Live Bee Relocations" — is how I discovered just what it was that bees did before beekeepers got to messing with them, and how I began to learn about natural wax. I did this kind of work for several years until it become too onerous to schedule, and I enjoyed every minute of it, even if it was often pretty stressful. It's rare that you will be confident that you have captured the queen in the process of removing bees from a building, so the likelihood of saving the colony is low, unless you have other resources to apply to the hive — like bars of open brood or a purchased queen if needed.

But if you are up for this kind of bee adventure — you'll certainly find that there's a special kind of rush involved, and to you I say "Enjoy. Bee Careful."[2]

Equipment and Supplies You Will Need

A Top Bar Hive—Assembled and Appropriately Placed

Location

If the climate in your location is snowy and cold in the winter, be sure to place your top bar hive in a sunny location — as close to full sun as you can. However, if you live in an area with very high temperatures throughout the summer, you should place the hive in an area that gets dappled sunlight or even some periods of shade throughout the day.

Direction

Orient your hive in such a way that there is no prevailing wind blowing directly into the entrance. Most often, facing south or southwest will accomplish this. However, if that is not the case in your location — by all means, change the direction that the hive faces!

Level

It's important that your top bar hive be as level as possible. The natural comb that the bees build inside the hive will be *plumb*. Plumb means straight up and down, just as if you'd hung a plumb bob down from your top bar. If the hive is not level, then you may find that the comb is attached to the bar in such a way that you are unable to remove the comb in order to inspect it.

Be sure that the hive is level, especially from end to end.
Credit: Christy Hemenway.

A Feeder

You will need a method of feeding your bees. A Gold Star feeder kit consists of a wooden tray and a one-quart mason jar with small holes punched in the lid. The tray is designed with wooden blocks that hold the jar high enough that bees can get under it, and low enough so that the jar still fits beneath the top bars so that they can all be installed above the feeder.

Any jar with small holes punched in the lid can be used to make a *feeder* in a pinch. Support the jar at least ⅜ inch above the floor of the hive so that the bees can get underneath it to reach the holes. Another inexpensive method of providing sugar syrup to your bees is to fill a sturdy plastic bag about half full, rest it in a flat container, then set the container on the floor of the hive and cut several two-inch slits in the top. This allows syrup to ooze out the slits so that the bees can reach it.

Be alert when using any sort of feeder though, as it can leak, making it look as if your bees are taking a lot more feed than they are, as well as making a sticky mess and attracting ants and other critters who come after the sugar.

Sugar syrup feeder installed.
Credit: Christy Hemenway.

Why Feed Sugar Syrup?

In the ideal world, an established colony will eat its own honey and shouldn't need to be fed. The 1:1 sugar syrup is for spring emergency feeding and in situations such as brand new hives where the bees have no honey stores. When you receive a package of bees to start a new colony in the spring, hopefully it is warm enough (consistently around 50°F) and there is enough forage available for the bees to start a new season of beekeeping. But the bees need carbohydrates so they can get started making wax. So, sugar syrup is intended to be used at the beginning of the season until the bees can begin to store their own healthy honey.

Feeding a New Colony

While sugar syrup may not be the best food for bees, it beats the alternative: starvation. So until they can get started building wax and successfully foraging on plants, please feed your bees! Feeding your bees sugar syrup is a stopgap method. As top bar beekeepers, you are probably not removing all of your bees' honey each year and then replacing that good, nutritious food with sugar syrup. I suspect that you are only feeding them when necessary—such as when they arrive in the package, or when there is an emergency situation like a dearth or a drought, or if they come up short in the fall. And that's as it should be. But don't get so caught up in being "all natural" that you don't feed, and wind up starving your bees.

Making Sugar Syrup

"A pint's a pound the world around…" You won't keep bees for long before someone quotes that old adage to you. What does it mean? It's meant to remind us that a pint of water weighs approximately one pound.

Sugar syrup, also called simple syrup, is made by dissolving granulated white cane sugar in warm water. The ratios are given as sugar:water.

In spring the recommended ratio is 1:1 for *spring start-up* feeding. In kitchen terms, that means five pints (ten cups) of hot water combined with five pounds of sugar. This will make a little more than four quarts of 1:1 syrup.

In fall, the recommended ratio is 2:1. That's two parts of sugar to one part of water. So in the fall, heat five pints (ten cups) of water, then stir in ten pounds of sugar.

Voilà—bee junk food!

Sugar syrup can also be made using various tonic teas as the water base, and with a touch of lemongrass essential oil added as well, which works to have a strengthening effect on the bees. Lemon juice can be added as a preservative. Here is an example of a Bee Tea Recipe:

> Prepare a mild tea by pouring boiling water over a mixture of chamomile, yarrow and dandelion flowers, leaves (and stems) of peppermint, rue, hyssop, horse tail, stinging nettle and thyme.... These teas should not be overpowering. You should be able to drink them yourself.[3]

Why Not Honey?

I am often asked "What about feeding the bees honey?" Well, as counterintuitive as it seems, feeding bees honey is a potentially harmful idea. This is assuming that you do not have your own natural, treatment-free honey available to feed them—or honey from a source that you trust absolutely, honey that you know comes from bees that were not treated with antibiotics. Because if you feed bees honey that was made by bees that were treated with antibiotics, it is possible that that honey will contain the spores that carry American Foulbrood—a devastating and highly contagious brood disease (see Chapter 9). And if you are a treatment-free beekeeper, your bees can contract American Foulbrood from that honey because your bees have not been treated with those same antibiotics. So in the absence of your own honey or honey that you can trust to be from an antibiotic-free source—please resort to good old "white death" sugar syrup. It's the closest we humans can get to imitating the nectar of flowers.

Some Important Notes about Feeding Bees

- Don't boil the sugar mixture—this caramelizes the sugar and creates solids that the bees can't digest and can give them bee diarrhea.
- For the same reason, do not use brown sugar or molasses as bee food.
- Please avoid feeding your bees high fructose corn syrup, as it's been found to contain elements that are harmful to bees, and the average ear of US-grown corn, the source of high fructose corn syrup, very likely contains three different systemic insecticides.[4]

Protective Bee Gear

A Bee Jacket and/or Veil

A bee veil is a crucial piece of protective bee gear. Most importantly, it protects your head and face. There are good reasons to protect your head from bees! One is hair. If you look at a bee's wings, you'll notice that they are shaped like a V. This means that if bees get into your hair, they are pretty much stuck, as they can't back out. They can only move forward — and the bad news about that is that it's your head at the end of their trajectory!

Getting stung in the face or head can be traumatic. A bee sting around your lips and eyes tends to swell more and to last somewhat longer than stings on other parts of your body. So I think it's a very good idea to protect your face and head when working with bees. I also think it's a good idea to wear a helmet when I ride my motorcycle — there's just no good reason not to, but many good reasons to!

There are plenty of options in the veil department. There is the veil and pith helmet combination, in which the veil drops over a separate helmet and has several yards of rope attached to it, which you tie around your body. There is also the complete suit, which looks like a pair of coveralls, with a veil attached by a zipper. This can be a great confidence-builder for the novice beekeeper. Its only drawback from my point of view is that you need to remove your shoes to put it on. Not a problem — just a little inconvenient.

I'm particularly fond of the bee jacket with a zip-on veil that you will see in photographs in this book. This builds in some body protection, and such jackets generally have elastic at both the wrists and at the hip, which is good for preventing bees from getting up inside the jacket. Yet you can shrug in and out of the jacket easily, and being able to wear it over a tank top or something very cool and lightweight in the summer is an advantage on a hot day.

Both jacket and suit can easily be washed, and although it's not recommended in most of the washing instructions, I have been known to toss my jacket — with veil still attached — into the wash all together with no problem. If you have family members who do not work with your bees, it's a smart idea to wash your bee gear fairly regularly, removing any dried venom from stings that may be present on the fabric. I have heard stories about beekeeper's family members who have never been stung—yet they develop a serious allergic reaction from being exposed to dried bee venom that gets into the air from the protective clothing of the member of their family who is a beekeeper.

A Pair of Bee Gloves

A pair of gloves is another crucial piece of beekeeping protective gear. Sticking your hands into a box filled with stinging insects is not likely to have a calming influence on the new beekeeper. Of course, gloves also make you slightly clumsier and affect your ability to feel what is happening when you are handling the comb in your hive. So choose gloves that will impede your sense of touch as little as possible. Goatskin is well-known as one of the most flexible leathers for gloves. Some people wear rubber or nitrile gloves, which allow for better agility and more tactile sensing ability. And when you've gained a bit of experience and confidence, it's likely that you will not feel that the gloves are required equipment; you may sometimes find it easier to work without them altogether.

Bee Logbook

I can't tell you how helpful it is to be able to look back and pinpoint the exact date I saw something in my hive. There is an inspection form in Appendix A that you can copy and use to record what you saw where in the hive; that, combined with the date, will help you to learn and keep track of what bees do when. This is the same inspection form used throughout this book. But however you do it—please keep a record!

Miscellaneous Supplies

- duct tape—use for taping any openings in your clothing or protective gear
- rubbing alcohol—use for splashing on bee stings to eliminate alarm pheromones and deter other bees from stinging

- spray bottle—while you are making the sugar syrup for your feeder, put about ¼ cup of it into a clean spray bottle. You can use this to lightly spray the bees when you are ready to install the bees into your hive.

Preparing the Hive for Bees

Setting up a Center Side Entrance Hive to Install the Bees

Picture your center side entrance top bar hive from above. It has 30 top bars. It has two follower boards. It has ten spacers, which you won't use until later in the season. In a center side entrance hive, the colony is installed in the center one third of the hive, aligned with the entrance holes and contained

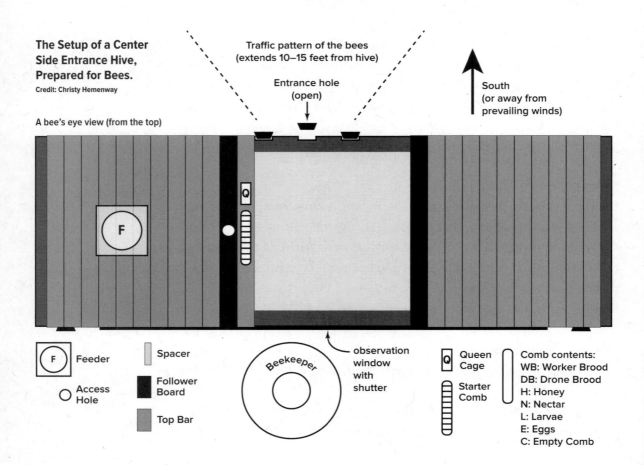

The Setup of a Center Side Entrance Hive, Prepared for Bees.
Credit: Christy Hemenway

A bee's eye view (from the top)

Traffic pattern of the bees
(extends 10–15 feet from hive)

Entrance hole
(open)

South
(or away from
prevailing winds)

F	Feeder		Q	Queen Cage
O	Access Hole			Starter Comb
	Spacer		
	Follower Board		
	Top Bar		

Beekeeper

observation
window
with
shutter

Comb contents:
WB: Worker Brood
DB: Drone Brood
H: Honey
N: Nectar
L: Larvae
E: Eggs
C: Empty Comb

between two movable walls, or *follower boards*. You will use the follower boards to divide your hive into three equal sections—each containing ten bars apiece.

One of the end sections, let's say it's the left side when you are standing behind the hive, will become the feeder section. It's not crucial which end the feeder goes in—but the follower board with the access hole must be facing that section to allow the bees to reach the feeder.

Set the feeder tray into the left-hand one third section. Turn the mason jar filled with syrup upside down over the ground outside the hive, and wait a few moments until it stops dripping. Then set the jar in the hive. Be sure it's supported in such a way that the bees can get under the jar to reach the feed through the holes in the lid. Install ten top bars over the feeder jar and then check to be sure that the follower board, which comes next, is indeed the one with the access hole in it. Set that follower board in place.

Now move to the right side of the hive and put ten bars across the top of that one third of the hive. Then put the follower board without the hole at the end of the ten bars. This one third, on the right side, will be the space your colony will expand into as the season progresses.

The Bee Bowl

Now you have the two end sections of the hive covered with ten bars each, and enclosed by the two follower boards—and an empty space, ten bars wide, in the middle. I call this the *bee bowl*. You are just about ready to install your bees into this open space.

Now, take a moment to apply your thinking beekeeper judgment. If you are installing a huge swarm, you might need to make this middle section a bit bigger—perhaps about 12 bars wide. Or, if your weather is cold and you are preparing to install a three pound package of bees, you might choose to reduce the bee bowl to about eight bars.

Entrances

One of the center front entrance holes should be open. The others should be closed.

The bee bowl, ready to receive bees (left), with bees installed (right).
Credit: Christy Hemenway.

Bottom Board

Your bottom board, if it's adjustable, should be closed if nighttime temperatures are cool, say below 50°F. It should be open if it's very warm out and nighttime temperatures are consistently above 50°F.

Setting up an End Entrance Hive to Install the Bees

In an end entrance hive, there will be only one movable wall, or follower board — the bees will be installed between the follower board and the front wall of the hive itself. The bees begin building comb near the entrance, and bars are added into the bee's space from the back of the hive as the colony expands.

If you plan to use an interior feeder, your follower board will need to have an access hole, or you may use a Boardman feeder, which goes into the long slot entrance of an end entrance hive with a landing board.

As you picture an end entrance hive from above, the bee bowl will be about ten bars of open space at the entrance end of the hive. Next will be the follower board, then the rest of the hive, covered by top bars.

Installing the Bees

Get Yourself Ready

Put your bee gear on. If you're using a tie-on helmet and veil, be sure to tie it securely so that the veil stays closely anchored to your body. If you've got a bee suit or a bee jacket, be sure that all the zippers are zipped; use the Velcro where all the zippers come together under your chin. The Velcro is there to prevent any stray bees from getting inside your veil.

Wear closed-toe shoes or boots (not open sandals or flip-flops). If you don't have a full bee suit, tuck your long pants into your boots, or into your socks.

Put on your protective gloves last. You won't really need to have your bee gloves on until you've actually removed the lid and the queen cage. Gloves make you a bit clumsy, but do what's most comfortable for you.

Getting the Bees Ready

Installing package bees is the most complex way to get bees into a hive. Opening the package, removing the queen cage and installing it in the hive, removing the feeder can and then *bonking* the bees are all specific to a package installation.

In the case of a swarm, you have none of that — you simply pour the swarm, queen included, into the bee bowl, described above.

With a split, you simply lift the top bars from the existing hive and place the bees into the bee bowl of the new hive.

This next section, with all its detail, is specific to preparing a package of bees for installation.

Helpful Items for Installing a Package

- screwdriver or putty knife — use for prying open the box (the package)
- pushpins — use for attaching the queen cage to a top bar in the hive
- sheetrock screw — use for removing the cork from the exit hole in the queen cage
- a hammer and a small nail — in case you need to pop a hole through something metal in order to hang the queen cage
- a spray bottle with a bit of sugar syrup in it

Get your package of bees and place the box on or near the hive. You are ready to begin!

If the weather is warm enough, say 60°F, you may choose to spray the bees lightly with the bottle of sugar syrup. If it's cold, don't do this — as you don't want your cold bees to also be wet. But if it's warm out, spraying makes the bees slightly sticky, which accomplishes several things. First it distracts them from you, which you want. Second it gives them something to do: They will be busy cleaning themselves up after you spray them. And third, it makes it more difficult for them to fly. And since you'd like them to go inside the hive cavity and mostly stay there until you get the top bars over them and the roof on, this is a pretty good method of getting what you want. Do not drench them — a light spraying is all that is needed.

> If it is cooler than 60°F don't spray your bees at all. Cold and wet is a disastrous combination for your bees.

Getting the Queen Cage out of the Package

Carefully pry the lid from the top of the package of bees. The screwdriver or putty knife will come in handy to do this. Your first goal is to remove the queen cage from the package. The queen cage sometimes lifts right out of a keyhole slot next to the can, or you may find that your package is constructed in such a way that you must remove the feeder can before you can get the queen out. Take your time, look at the way the package is constructed and act in accord. There are several variations on the bee package box, and different ways for the queen cage and feed can to be installed.

Remove the feeder can by pressing on one side and catching the opposite edge.
Credit: Christy Hemenway.

Getting the Feeder Can out of the Package

The round hole that the can of feed sits in is usually a fairly close fit. The best method I have found for getting it out is to push down on one side of the can, then catch hold of the raised edge and lift it more or less straight out. Your screwdriver or putty knife may be helpful when it comes to catching the edge and lifting the can. Be patient and work carefully. There's no real hurry at this point. The bees are still contained even though the lid to the package has been removed.

Once you've got the queen cage free and the feeder can out, the bees can now get out of the package. Carefully set the lid back over the hole to keep the bees inside. If the lid has nails or staples in it, the ones that were holding it to the package, simply turn it upside down so that they point up. Don't get stuck by them!

Preparing the Queen Cage

The queen bee will be confined in a small cage. The cage will have one, possibly two exits. One exit of the queen cage will be plugged by sugar candy. There will be a cover of some kind over this candy plug; you must remove this cover so that the bees can get to the candy plug. If it is a metal flap, you can simply bend it back. If it is a cork, you can use a sheetrock screw to remove it: Gently poke the sheetrock screw into the cork, twist and pull it out. Sometimes there is a cork in both ends of the queen cage — be sure that the cork you remove is actually covering a candy plug, and that it is not the only thing keeping the queen inside the cage! Ideally, you want the bees to take a day or two to get through the candy in order to release the queen slowly.

Hanging the Queen Cage

You will probably be able to use whatever fastener was holding the queen cage in the bee package to attach the queen cage to a top bar. Some queen cages come with thin metal strips attached

> ## Time-Release Queen
>
> The purpose of the queen cage is to keep the queen separated from the other bees while they acclimate to her pheromones. But once the bees are in the hive, she will need to be released from her cage. The other bees will go to work to eat the exposed candy in the exit hole and release the queen over the course of the next several days.

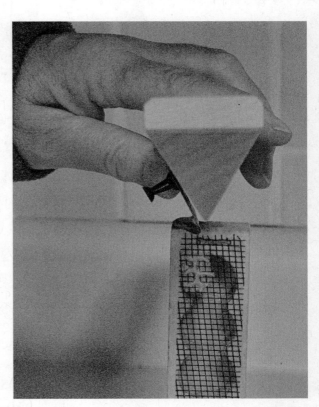

Be sure to have the queen cage hang directly beneath the top bar.
Credit: Christy Hemenway.

Gently pour the bees into the bee bowl in the hive.
Credit: Gold Star Honeybees®.

that hold them in the package, some with plastic strips. Some have a metal disk. Using whatever method seems reasonable — along with a pushpin or perhaps a rubber band — attach the queen cage firmly to one of the top bars. Strive to have the cage hang directly below and in line with the bar, as you don't want the bees to start to make crooked wax combs by going around an awkwardly placed cage.

Bonking and Pouring the Bees into the Hive

Pick up the package, holding the lid over the hole, and firmly tap one corner of the package on the ground. This is humorously referred to as *bonking the bees*. A bonk is harder than a tap, but softer than a slam. Be serious about it, but not overenthusiastic!

This sharp motion will cause the cluster of bees to fall into a heap inside the package. Take the lid away from the hole, and gently pour and shake the bees into the bee bowl — that empty portion of the hive that you prepared. Thump on the box a little bit to help knock the majority of the bees out of the box and into the hive. When most of them are in the hive, you can set the box with its stragglers on the ground beneath the hive. They will find their way in eventually.

Finishing the Installation

Top Bars

Pick up the remaining top bars and carefully place them over the cavity into which you poured the bees. Try not to crush any bees, but make sure the bars touch one another. If there is extra space — and sometimes there is since wood expands and contracts due to heat and moisture — then make sure the extra space is outside of the space where the bees are installed, and fill it in with some spacers.

Be sure that the bees are contained by the follower board or boards, and a solid ceiling of top bars. If the bees can access the attic space above the bars because no follower board is preventing it, or because all the top bars are not in place, they may begin to build comb down from the inside of the roof of the hive, and this you definitely do not want!

Roof

Next, set the roof of the hive in place.

I like to run a nylon strap with a metal thumb clip around the hive and through a dog "corkscrew" — a tethering device that screws into the ground. This keeps the lid secure and anchors everything to the ground (see page 125 in Chapter 7).

In a Gold Star hive, the gable roof is designed so that the bees can walk off the ends of the top bars and get out of the roof area—so there is no need to brush them off. They will find their way out and into the hive.

Breathe!

Now you can step back and take a deep breath. If this is your first hive — congratulations! You are now a beekeeper.

CHAPTER 6

Inspections

In this chapter you will find detailed instructions and clear answers to the questions that most people have about how to manage a top bar hive.

But far more important than any rote checklist of tasks, or calendar of when to do them, is for you to understand why you are doing what you are doing. I really want you to get that — the why of things — so that you are not dependent on lists and calendars. When you understand why you are doing something, then you will be able to adjust the how, what, when and where of your hive management in the event that you find yourself in different circumstances — whether that is a different location, a different growing season or different set of weather circumstances.

And I think you want to know the why too — that's probably the reason that a book entitled *The Thinking Beekeeper* made sense to you — because you wanted more than just a mindless list.

In Chapter 3, I talked about bee biology and the bee math that goes with it. Here's a review of the bee math as it relates to inspecting your top bar hives. To the right is a gestation cycle timetable for queen, worker and drone bees.

Gestation Cycle of Queen, Worker and Drone Bees

Caste	Egg	Larva	Pupa	Total Gestation
Queen	3.5	4.5	8	16
Worker	3.5	5.5	11	20
Drone	3.5	6.5	14	24

Credit: Christy Hemenway.

Why Do Hive Inspections?

Beekeepers inspect hives to check on the health and progress of the colony. If there is a disease problem, it is important to catch that in the early stages, before it is devastating to that colony or infects other hives. If there is another issue, such as queenlessness or a laying worker, it is important to catch that as early as possible too.

Wax and Inspections

When starting a new colony in a top bar hive, you are the shepherd of your bees' wax production. Movable comb is a requirement of managed beekeeping. In order to follow the colony's progress and to catch any issues early, you absolutely must be able to remove the combs to inspect for disease. Do not kid yourself into thinking that you are doing your bees a favor by "leaving them alone" in a top bar hive. If there was a problem and you could act to help or to correct the problem, wouldn't you want to do that? If the comb is not movable, you will not be able to help your bees without destroying the hive.

Not to mention that movable comb is required by law in most places — for all of the very good reasons stated above.

So your first job is to maintain movable, inspectable combs by preventing the bees from building *cross-comb*. That's the term for what happens when the bees begin to draw their wax combs across the top bars instead of in line with them, effectively locking the bars together and making it impossible to inspect the hive.

Be aware that a wax problem never gets better — it only gets worse! And it can get worse very fast, so you need to be very vigilant in the beginning. The bees will not suddenly "get it" and straighten out their comb-building habits midway through the hive; they will simply repeat the pattern that they have begun. This is why you want the wax to be straight on the top bars from day one. If the bees begin to cross-comb you want to catch and correct that problem very early.

How to Correct a Cross-Comb Wax Problem

What do you do if your bees don't get it, and they don't draw their wax in line with the comb guide?

Should you discover that your bees have begun to build their honeycomb crooked off the bar, it is imperative that this problem be corrected right away. The bees will maintain bee space — that all-important ⅜ of an inch — between their combs, causing any curves or other deviations from the parallel top bars to be repeated and even exacerbated with each bar of comb they create. Because this often means that the bees are building across multiple bars, this can result in a hive that is completely cross-combed, or locked together, effectively creating a fixed comb hive, which you definitely do not want!

Here are some ways to prevent or correct the problem:

- To encourage the building of straight comb, you can insert a blank bar between two bars of straight comb that the bees have already drawn, or between a straight bar of drawn comb and a follower board. The bees must maintain bee space on either side of the new comb they are drawing, so voilà—straight comb!
- If only a part of the wax on a bar is crooked, sometimes a correction can be made by cutting the offending piece and attempting to gently push it back in line with its original bar.
- Another option is to look down at the bar from above and simply cut off only the piece that is visibly out of alignment. The bar can then be replaced into the hive, and the bees will repair that portion of the comb that was removed.
- If the comb itself is straight, but it's not attached in alignment with the bar, you can remove it completely and create a sling, using fabric strips, or pieces of ribbon, tacked to the top bar and running beneath the piece of comb. Use this sling to hold the comb as close to the bar as possible, and the bees will reattach it to the bar. In time, they will probably remove the fabric strips as well. You can insert a blank bar between the follower board and this repaired bar—and get another straight bar between them.
- NOTE: When wax honeycomb is new, it very soft and quite fragile. Be very gentle when inspecting and manipulating the comb!

To Wax or Not to Wax—What about the Top Bars?

When Gold Star Honeybees® was founded, it was typical for top bar beekeepers to paint or to rub beeswax on top bars. This was believed to serve two purposes: 1) to act as a guide, showing the bees where to build their comb and 2) to put the smell of beeswax in the hive.

Over time though, I've come to realize that no matter how we humans attach beeswax to the top bars, the bees can attach it better! In fact, there have been so many reports of entire combs collapsing off pre-waxed top bars that I now recommend beekeepers put nothing at all on the points of the top bars. In addition, the bees don't necessarily follow any such painted-on beeswax as a guide, so applying wax is most often a waste of effort. Clean, dry wood and a long, beveled guide on your top bar are best.

Anti-Absconding Tricks

But what about having the smell of beeswax in the hive? This, I've discovered, is a valid concern. There have been many instances of bees *absconding* from a brand new, clean, top bar hive that contains nothing that smells like home to the bees. The best solution I have found for this has been to put in a piece of brood comb from an existing hive — in fact, the darker and smellier the brood comb, the more the bees seem to like it.

In hives where the beekeeper hung a piece of brood comb, parallel to and just below a top bar, and applied several drops of lemongrass essential oil to the interior wood of the hive body, the number of colonies that chose to abscond dropped to zero.

Handling Honeycomb

You will want to train yourself thoroughly and well on this aspect of managing a top bar hive. It's not that it's difficult — it's just different.

There are two things to keep in mind when you lift a top bar from the hive:

1. **Always keep the comb itself vertical.** The comb can be held upside down, or even extending out to the left or right from a top bar being held vertically — but the edges of the comb must always be straight up and down, in line with gravity. Never hold the bar so that the comb extends out horizontally (flat).

   Why? Because the weight of a full bar of comb filled with honey (which is quite heavy) or even with brood (which is somewhat lighter) is enough to break the comb from the bar in that position.

Always keep the comb in line with gravity— never flat-wise—to avoid breakage.
Credit: Tom Frields.

2. **Put the bar back into the hive oriented the way it was when you removed it.** One way to be certain that you do this is to mark one end of each of the top bars. This way you know you are putting the bar back into the hive in the same orientation as when you removed it. Why does this matter? Because the bees configure their brood nest as they go, each bar fitting next to its neighbor like a jigsaw puzzle. Turning the bars will disrupt this puzzle, and often the combs will not even fit properly next to each other.

Hive Tools for a Top Bar Hive

The most effective hive tool I have found has been a long straight-bladed knife, with a fairly hefty handle. A roast beef carving knife is a good example. I prefer a straight, smooth-edged blade over a serrated edge as it is less likely to tear at the comb, cause damage or pull the comb from the bar.

Using a hive tool to carefully separate the top bars.
Credit: Gold Star Honeybees®.

A hive tool can be used in many ways when working in your top bar hive.

1. Its blade can be inserted between top bars and gently twisted in order to separate the bars before lifting them from the hive.
2. The handle is particularly useful when cutting away any attachments that the bees may have made to the inside of the hive. I do this with what I call the *windshield wiper technique*. Carefully insert the blade straight down alongside the inside wall of the hive next to the comb with the attachment. Let the handle come to rest on the top edge of the hive body. Then, using the handle as a fulcrum, swivel the blade up — toward the wax attachment. This cuts the attachment wax from the bottom up and prevents any downward pressure on the comb.
3. The hive tool can also be inserted between bars when there is cross-comb at the top: Get just enough space available to insert the knife and cut along the diagonal of the bar.

Replacing the Top Bars When there are Bees in the Way

Bees often get between the top bars and are in the way when you go to replace the bar. The best way I have found to get the bars back in place without crushing bees is to gently nudge the bees with the side of the top bar, lightly bumping and then releasing them. They respond to the gentle pressure of the bar, and when the pressure is relieved, they will walk out of the way. When you release the pressure do not move the bar very far — only far enough that the bees are released and can then safely walk out of the space.

This is when I usually find myself talking to the bees in a singsong voice and saying "Excuse me, ladies!"

The design of the top bars themselves can help to crush fewer bees. There is no need for bars to be thick vertically, as they are not supporting any large amount of weight. In fact the less surface area there is to the sides of the top bars, the less area there is to crush the bees. So a top bar with ⅜-inch vertical sides makes your life much easier when it comes to replacing the bars in the hive than a thicker bar. The figure on page 51 in Chapter 4 illustrates this difference.

Keeping the Bars Touching While Inspecting

Because the bees do tend to get in the way when you try to replace the bars close together, there is a tendency to set the bars down with space between them and then plan to close them up tighter later. But I suggest that you take the time to place the bars tightly back together each time you inspect one bar and return it to the hive. This keeps the bees in an enclosed space even while you are in the midst of an inspection; which helps to keep the bees calmer.

It also means that the working space you opened up in order to get into the hive is maintained: For instance, if you have two bars' worth of open working space, you can't afford to give up that space by leaving space between each of the bars you have inspected.

To Smoke or not to Smoke

I confess that I have never, in my years as a top bar beekeeper, used a smoker when inspecting a top bar hive. There have been a couple of times when the bees were grouchy enough that it may have made the inspection easier for me — but since I had already begun and had the hive open, I wasn't going to go off to get a smoker in midstream and try to get it lit with thousands of aggravated bees flying about. Bees in top bar hives just seem to be less likely to be upset by the inspection process than those in a conventional hive. I suspect that an important reason for this lies in the difference between the basic construction of the two hive types.

Why Beekeepers Use Smoke

There are a number of reasons usually given for smoking the bees when working a beehive. Probably the most common is that it calms the bees. Another explanation sometimes given is that smoking makes them believe that their hive is on fire, and so they gorge on honey—which makes them suddenly very busy, and once full of honey, very placid.

It doesn't make much sense that when the bees' environment suddenly fills with smoke, they somehow magically become calm.

However, many things in the bees' world—from their ability to sense the health of their queen to being able to identify the source of a threat—hinge upon their ability to smell or sense the source of phero- mones. There are multiple types of alarm pheromone—including one that signals a general, low level of irritation and another that signals a very high level of alarm, which is emitted when a bee stings.

So perhaps it more logical to think that when the beehive is filled with smoke, these pheromones are masked, and the bees' ability to identify the source of a threat be- comes confused. This would make them less able to identify the beekeeper as a threat, and perhaps less likely to sting said beekeeper.

When the top covers of a Langstroth hive are re- moved, the entire population of the hive becomes aware of the intrusion all at once. All of the climate and humidity they have been working to main- tain in the interior of the hive—whether cooler or warmer than the outside world—disappears in an instant. That aggravates the entire colony, and you can often see and feel their irritation when they all come to the top of the frames and look up at you, buzzing loudly!

In a top bar hive, when the roof is removed, none of the interior of the colony is exposed at first. Even when one or two top bars are removed to get enough room to work, only one side of one bar is initially exposed. Then as you go through the hive, bar by bar, if you replace each bar firmly against its neighbor, this keeps the disturbance and stress dur- ing an inspection at a much lower level. So smoking the hive often just doesn't seem to be necessary.

Smoke does seem to work most of the time, pro- vided you can get and keep the smoker lit (grin)— so if you feel your confidence will be increased with the use of smoke, by all means, feel free!

Keep a Bee Log

What you see during any given hive inspection can be helpful in troubleshooting problems, anticipat- ing events, scheduling subsequent inspections and preparing for an increase in the size of your apiary. This is the best reason I can think of for keeping a bee log of some kind. Recording the dates of your inspections will go a long way toward helping you discover whether you have a healthy hive or an issue with your queen; a thriving hive that is preparing to swarm; or some other event where you might want to intervene. Being able to look at your inspection records and see when you saw what is just really helpful.

A three-ring binder style notebook is one good choice, with dividers, making a section for each hive. Appendix A includes a blank master copy of the diagram used throughout this book to depict the progress of the hive. Feel free to copy and use this to track your hive's progress; adding additional notes of your own will help you create a clear record of your beekeeping journey.

A piece of paper attached inside the lid of the hive can also work for recording purposes, and there is even online computer software available that can be downloaded to your smart phone.[1]

Other things to record in your bee log include

- the weather on the day you inspect: the current temperature, whether it is sunny, cloudy, still or breezy
- the weather between inspections: Has it been rainy? hot? dry? cold?
- what is currently blooming in your area? This will help you understand your bees' ability to bring in food stores
- the number of combs drawn, the number of bars they have in their space
- what you see in the comb: eggs, larvae, capped brood
- any signs of swarming: queen cups, swarm cells
- any signs of queen issues: lack of brood or supersedure cells

Always make a note concerning anything that just doesn't look right—or any signs of pests or disease. You will find that many of these will probably just be something you haven't seen before—but it's always worth having a note saying when you first saw them.

Start keeping track immediately when you set up the hive. Note the date that you installed the bees, and then note the date of each inspection and what you see. It's also smart to keep a day count—to track the number of days/weeks/months old your hive is.

If you do this, at the end of the season you will have a good record of the life of that particular colony of bees in that particular hive. You will be surprised at the differences between hives — so be sure to keep a log for each hive. It will be both fun and helpful to be able to look back and see what happened. You will definitely learn a lot.

How to Inspect a Top Bar Hive

Any kind of inspection is going to disturb the bees, at least a little. So, don't do it lightly. Have a thoughtful plan for opening your hive. Generally speaking, you will want to inspect more frequently in the spring and less frequently as the season progresses.

So let's begin…

1. Walk out to the hive, and stand off to one side

Observe the hive entrances. Note the amount of activity and the general energy of the hive. Listen to the sound of the bees. This is a good indicator of how the bees are feeling. Loud, angry buzzing or frenetic, agitated behavior will tell you to be on the lookout for something wrong. Calm, quiet activity is a good sign of happy, healthy bees. Look at the bees' legs. Are they bringing in full pollen baskets? Are there lots of workers? Do you see any drones? Just get a general feel for how things are going.

2. Remove the shutter and look in the observation window

Many top bar hives have an observation window built into the side. This is a fantastic tool for getting a quick status check on your colony. The things that you can see in a window inspection include whether a feeder needs to be refilled and, if it is a new hive just building its first wax, you can tell whether your bees need more top bars to build comb on. This ability to peek through the window is less stressful on both you and your bees than a full bar-by-bar inspection, and it's rarely required that you suit up just to peek into the window.

However — don't fool yourself into thinking that a window inspection is adequate for truly evaluating the status of the colony. Top bar hive beekeeping, with its emphasis on natural methods, has a bit of a reputation for being "leave 'em alone beekeeping." But please don't take this to mean that you should not inspect. You really do need to inspect — gently, but thoroughly and thoughtfully.

If your hive includes an observation window, be sure that it has a shutter that you can close over it. Bees prefer to build their colonies in a dark, sheltered, enclosed area. Having light entering the hive at all times via an uncovered observation window is very likely to cause your bees to *abscond*.

3. Check the debris on the bottom board

Many top bar hives also have a screened bottom with a removable or adjustable bottom board. This is very useful for monitoring for varroa mites by lowering the bottom board and installing a *sticky board*. The location of the debris on the bottom board will also give you an indication of where inside the hive things are happening with the colony, so give it a glance before you open the hive. Or if you have a screened bottom on your top bar hive and your bottom board is removed, then you have a unique opportunity to see what your bees are doing — by looking up at them from underneath.

4. Inspect bar by bar

This is the fundamental in-depth portion of the thorough inspection you need to do in order to see what is actually happening inside the hive. It is imperative that you inspect each bar, looking for the activities the bees are engaged in and learning to comprehend what you are seeing. Inspecting is less about manipulating the colony than it is about your peace of mind.

This thorough sort of inspection should be done after donning your protective gear. Some beekeepers may tell you that they don't need to wear a veil when they work their bees — and indeed some days that may be true — but as a novice beekeeper, you are likely to find that your confidence level is much higher when you feel that you are armored against being stung. A sting can be very startling — and could make you drop a bar, break a comb or could even cause you to accidentally crush your queen if you began to rush. As you gain confidence, you will be better able to decide whether you need to wear a veil and gloves, but starting out it just makes sense.

What Things Look Like

I remember going to beekeeping school. There was a lot of information presented, lots of equipment to learn about and lots of requirements about when to do what and at what temperature and which chemical went where when. I knew when I finished bee school that I was supposed to inspect my hives. But I had not the foggiest notion what I was inspecting for. I could not have told you a larva from a pupa from a drone cell from a queen cell.

Because I know that novice beekeepers struggle with what things look like and what they should expect to find during an inspection, I will do my best to give you pictures and drawings to make caring for your bees easier.

What follows next is a very general time line describing what you may find during an inspection, based roughly on the age of the hive. You may need to compensate for the variables, like size and weather, that affect each and every hive. You will note that the range of days shown for each illustration frequently overlap. Partly this is a demonstration of what you will see in reality, especially if you have more than one hive, and partly this is my attempt to keep you from falling into the habit of relying upon a calendar-based checklist.

The imprecision of the chronological progression of a beehive through the seasons is another good reason to keep a bee log — especially as a novice starting a new colony. When you know the day the bees were hived, you can at least anticipate what to look for, based on the age of the hive.

This description begins with Day One, having installed bees into the top bar hive as described in detail in Chapter 5. It follows the colony through the season to a full-fledged thriving colony of healthy bees and combs full of healthy honey, and then through winter shutdown procedures.

The hive diagram drawings that follow detail the progress of a center side entrance style of top bar hive. An end entrance hive progresses in the same way, with the difference that the left follower board in these drawings would actually be the front of the hive. In an end entrance hive, all the expansion will occur from the front of the hive toward the back (left to right in these diagrams), and there will be no need for a mid-season shift.

Day One: Hiving the Bees

This is what things look like on Day One in a center side entrance hive. You have installed your bees — whether a swarm or a package, in a space of approximately ten bars. This assumes a relatively mild day, with temperatures at least high 50°F–low 60s°F. If it is colder out than that, install them on fewer bars — eight or even fewer — depending on the temperature. But be sure that you don't go so small that there isn't enough room for bees in the space.

You have hung your queen cage and a piece of starter comb off the first top bar — that is against the follower board with the access hole to the feeder. You've put a few drops of lemongrass essential oil on the inside of the hive body.

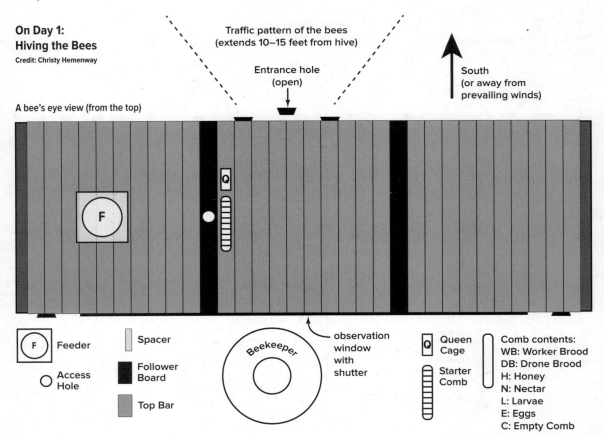

On Day 1:
Hiving the Bees
Credit: Christy Hemenway

A bee's eye view (from the top)

Traffic pattern of the bees
(extends 10–15 feet from hive)

Entrance hole
(open)

South
(or away from
prevailing winds)

F — Feeder

○ — Access Hole

▮ — Spacer

■ — Follower Board

▮ — Top Bar

Beekeeper

observation
window
with
shutter

Q — Queen Cage

▯ — Starter Comb

Comb contents:
WB: Worker Brood
DB: Drone Brood
H: Honey
N: Nectar
L: Larvae
E: Eggs
C: Empty Comb

There is fresh sugar syrup in the feeder.

In the case of an end entrance hive, the queen and starter comb are hung against the front of the hive, which can be pictured as the left-hand follower board.

Write this date in your bee log! This is Day One for purposes of doing Bee Math.

Days 3–10: Removing the Queen Cage

In three to seven days after hiving your bees, the hope is that the worker bees will have eaten their way through the candy plug in the queen cage, releasing the queen.

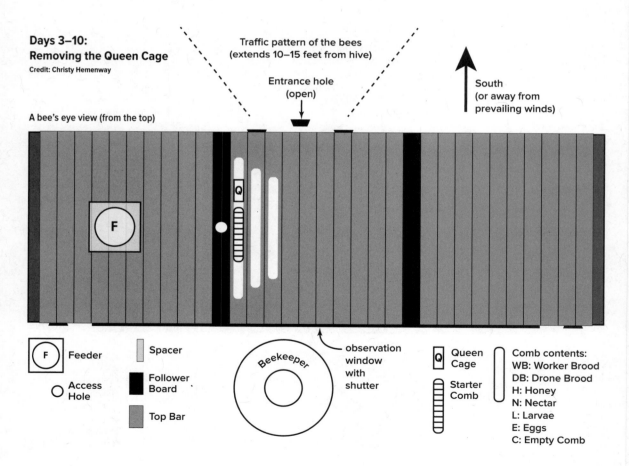

Days 3–10: Removing the Queen Cage
Credit: Christy Hemenway

Traffic pattern of the bees (extends 10–15 feet from hive)

Entrance hole (open)

South (or away from prevailing winds)

A bee's eye view (from the top)

F — Feeder
O — Access Hole
Spacer
Follower Board
Top Bar
Beekeeper
observation window with shutter
Q — Queen Cage
Starter Comb
Comb contents:
WB: Worker Brood
DB: Drone Brood
H: Honey
N: Nectar
L: Larvae
E: Eggs
C: Empty Comb

Carefully cut away any wax that the bees have built, then remove the queen cage.
Credit: Gold Star Honeybees®.

If the queen has been released (and you've got very good vision!) then you may see eggs in some of the cells. They will be white, shaped like a grain of rice, but about the size of a hyphen (-) and of course, they'll be very difficult to spot since the comb is white as well. Not to mention you are looking through your bee veil, and you probably aren't wearing your glasses. But if your queen was quick to be released and to begin laying you may even see some larvae already; larvae are much easier to see. So look for signs of the queen having been released—in the form of eggs or larvae.

If the queen is still in the cage on day seven, it is definitely time to take action to release her. Remember, the queen that came with a package of bees will have been raised in a queen breeding facility and will not have any prior connection to the worker bees in the package. The purpose of the queen cage was to protect the queen from the worker bees until they have become acclimated to her particular pheromone, but by this time you really want her to be out in the hive and laying eggs.

Check to see whether the queen has exited the cage. If she has, you can now remove the queen cage. If she has not, you can release her by carefully running a nail or screw through any candy that is still blocking the opening

A Dead Queen?

What if, when you check to see if the queen has been released, you find her dead in the cage? Don't panic! This does happens, and it can have more than one cause.

Scenario One

One likely scenario is that she died in the cage either while on her way to you or shortly after you hived the package. In this situation, with a brand new package, the bees have none of the necessary resources—in other words, there is no open brood containing young larvae in the hive—to make a replacement queen. So your next step is to either

1. procure another queen, or
2. procure a bar of comb containing open brood from another top bar hive beekeeper, and provide that to your bees so that they can make an emergency queen.

If you contact the apiary from which you purchased the bees, it is likely that they will be willing to replace a queen that died before getting out of the queen cage. It is worth asking about the apiary's queen replacement policy when you purchase a package of bees.

Scenario Two

The other possibility is that a loose queen bee was accidentally shaken into the package during the process of packaging the bees, and that that queen—or worker bees defending that queen—killed the caged queen. In this case, installing a new caged queen into the hive will probably just get her killed too.

So how can you tell if Scenario Two is your situation? Examine the combs very carefully, looking for eggs and/or larvae and/or a queen bee in the hive. If the queen was loose in the package, there was nothing to delay her starting to lay eggs—she only had to wait until the worker bees made a bit of comb for her to start laying—and in fact, she probably began laying in the starter comb, if you had that. So by day seven it is very likely that you will be able to see larvae, or at the very least, some eggs.

If that's the case, then you're all set, and you don't need to do anything more. The important thing is to look for signs of a existing queen before you spend money on a replacement queen who could be doomed from the start!

and letting her make her own way out. Don't poke her if she's still in there! Alternatively you can pry the staple from the front of the cage and fold back the screen to release her directly into the hive. Be very careful to hold the queen cage over, or even down into the hive as you do this: You don't want your queen to fly off and get lost, or to wind up on the ground, where she could easily be stepped on.

You should see the first few bright, fresh combs. The bars may not be completely filled out to the edges of the hive yet — but the bees should have started to build comb on several top bars.

The bees may have built some wax onto the queen cage. A knife with a pointed tip, like a steak knife, may be helpful here — carefully use the tip of the knife to cut any wax away from the cage.

Check for cross-comb problems. Remember, a wax problem never gets better — it only gets worse — and fast! Correct this immediately if you see it. (See "How to Correct a Cross-Comb Wax Problem" earlier in this chapter.)

Refill the feeder if necessary.

Remember to write these things in your bee log.

Days 10–20: Early Inspection

By day ten the queen should be out and starting to get busy laying. Your bees should have made a good beginning on building some honeycomb.

You may be seeing some larvae, and the starter comb is probably being incorporated into the comb that the bees are building.

Over the course of the next several inspections, you will see the bees continuing to build comb toward the right. There will be more and more larvae, and soon there will be capped worker brood.

In an end entrance hive, this growth will be occurring from the front of the hive toward the back.

Worker brood cappings are flat and look somewhat fibrous. Drone brood have raised, dome-like

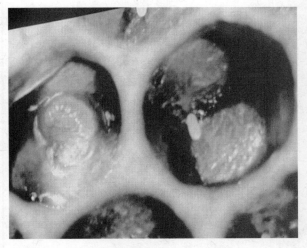

A honeybee larva (left), honeybee egg (right).
Credit: Christy Hemenway.

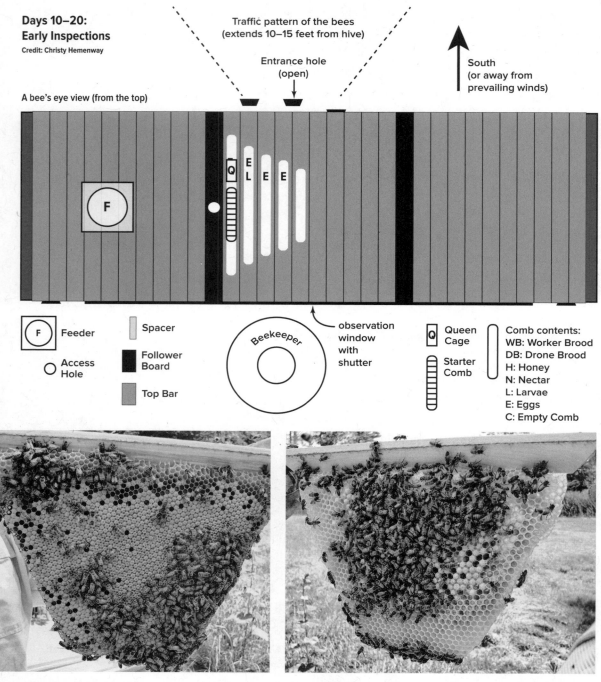

Days 10–20:
Early Inspections
Credit: Christy Hemenway

A bee's eye view (from the top)

Traffic pattern of the bees
(extends 10–15 feet from hive)

Entrance hole
(open)

South
(or away from
prevailing winds)

F

Q E E E
L

F Feeder

◯ Access
Hole

Spacer

Follower
Board

Top Bar

Beekeeper

observation
window
with
shutter

Q Queen
Cage

Starter
Comb

Comb contents:
WB: Worker Brood
DB: Drone Brood
H: Honey
N: Nectar
L: Larvae
E: Eggs
C: Empty Comb

Capped worker brood with its flat, fibrous cappings.
Credit: Jim Fowler.

Capped drone brood with its raised, dome-like cappings.
Credit: Jim Fowler.

cappings—they look a bit like puffed-grain cereal. It would be early for the queen to be laying drone brood—but it is pictured on the previous page for the sake of comparison.

Days 20–40: Hive Expansion

During this time the bees are building comb, the queen is laying eggs and you should find yourself with several attractive bars of brood comb. The space that the bees were originally hived in will begin to fill up, and soon it will be time to begin adding blank top bars in the direction the bees are building.

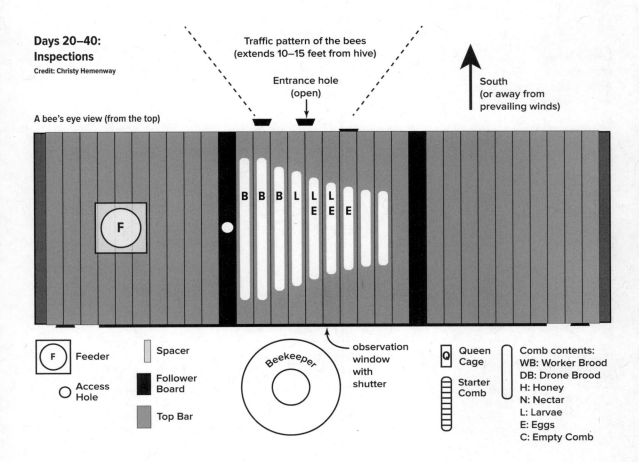

Days 20–40: Inspections
Credit: Christy Hemenway

Days 30–50: Adding Bars In One Direction

The beekeeper's task at this stage is to continue to act as a wax shepherd, preventing or correcting any cross-combing that may occur, gently detaching wax from the sides of the hive when doing inspections and adding bars into the bees' space, as needed, in the direction that the bees are building (in the case of our example, to the right).

You usually want to have at least two bars available for the bees to build comb on—so you will find yourself adding one to two bars at a time from the space to the right into the space between the follower boards.

This is a good time to do your first monitoring of your hive for varroa mites. It's important to be certain that the mite load in your hive is low

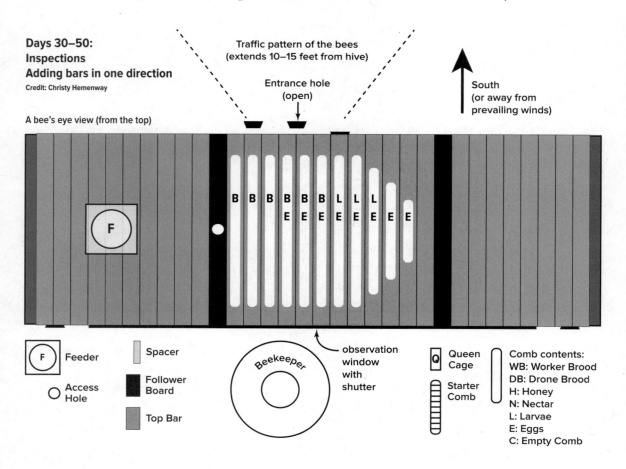

Days 30–50:
Inspections
Adding bars in one direction
Credit: Christy Hemenway

A bee's eye view (from the top)

Traffic pattern of the bees
(extends 10–15 feet from hive)

Entrance hole
(open)

South
(or away from
prevailing winds)

Feeder	Spacer	Beekeeper	observation window with shutter	Queen Cage	Comb contents:
Access Hole	Follower Board			Starter Comb	WB: Worker Brood DB: Drone Brood H: Honey N: Nectar L: Larvae E: Eggs C: Empty Comb
	Top Bar				

enough that no action needs to be taken. Because varroa mites can reproduce so quickly and are so talented at helping to spread disease, it's important to keep them under control. See Chapter 9, Varroa Mites, What Can You Do About it? for detailed instructions on monitoring and treating for mites.

Days 40–60: Continuing to Expand

In the diagram on page 104 the beekeeper has continued to add bars from the right-hand side of the hive into the space between the follower boards — moving the solid follower board further to the right, and giving the bees room to expand their brood nest — in one direction only. This expansion

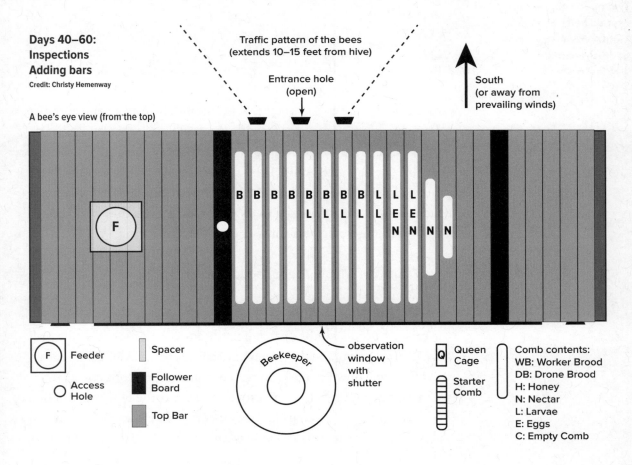

Days 40–60:
Inspections
Adding bars
Credit: Christy Hemenway

A bee's eye view (from the top)

Traffic pattern of the bees
(extends 10–15 feet from hive)

Entrance hole
(open)

South
(or away from
prevailing winds)

F

B B B B B B B
L L L L L L E
N N

B L L
L E E
N N N

observation
window
with
shutter

Beekeeper

F Feeder

Access
Hole

Spacer

Follower
Board

Top Bar

Q Queen
Cage

Starter
Comb

Comb contents:
WB: Worker Brood
DB: Drone Brood
H: Honey
N: Nectar
L: Larvae
E: Eggs
C: Empty Comb

Two queen cells—one capped (left), one open (right)—from a hive that has just swarmed.
Credit: Jim Fowler.

toward the right is the goal in a center side entrance hive. In an end entrance hive, this expansion is occurring from the front of the hive toward the back.

At this stage, you may see drone brood (see photo on page 102) as well as queen cups, or even queen cells. These are signs that the hive is considering swarming. However, just because you see a queen cup does not guarantee that the colony is truly planning to swarm — sometimes they build queen cups only to tear them down again as if they were practicing.

What to Do if Your Hive is Preparing to Swarm
Here are some options:
1. Open up the brood nest by adding single blank bars between some of the existing brood combs. Expanding the amount of room that the bees have to build comb, and the queen has available to lay in, often has the effect of slowing down or postponing the swarm impulse. Be conservative when doing this, since to some extent, it disrupts the bees' construction of the brood nest. I generally add no more than four bars in this way.
2. Split the hive. This means that you, or another top bar beekeeper nearby,

need to be ready to expand your apiary and have additional equipment ready to receive bees. If you split the hive before they swarm, you eliminate two worries: 1) that the bees will land so high above the ground that they are not retrievable and 2) that they will move into another inconvenient location (see #4 below).

3. Let them swarm. In this case, the colony swarms, and you then collect the swarm and hive them. This only works well if you happen to see them swarm, and if they land in a place where you can safely retrieve them.

4. Do nothing. This is always an option with bees. However—if they find any small hole or crevice in the wall or roofline or your house, or your neighbor's, and choose that as a cavity to move into, then you have a live bee relocation situation on your hands—and possibly some bad PR for your bees.

Please practice good bee citizenship, especially if you are keeping bees in an urban or suburban area. Preempting a swarm by splitting your hive when you see them preparing to reproduce combines the best and most practical aspects of bee-keeping. It works to support the bee's natural system by letting them begin the swarming process, and it also helps to prevent issues with neighbors who may not be as enamored of your bees as you are.

Days 60–80: Filling the Hive in the Original Direction

So you've been adding bars, moving the follower board to the right (or toward the back) of the hive. It's been a couple of months since Day One. When your hive has got to this size, you are probably starting to feel pretty good about the bees'

Tanging the Bees

Tanging (banging on a metal pot or can with a metal utensil) was something I mostly suspected of being an old wives' tale until one spring when I suddenly got three consecutive reports of it being extremely effective—not only in bringing bees down to a lower altitude, but in causing four clumps of swarming bees to coalesce into two groups, and then finally into one.[2]

The background of tanging is sketchy. Sometimes it is described as a way of following the bees in order to claim the swarm. In other stories the purpose of the banging of metal on metal is to induce the bees to coalesce into a cluster in a lower location, making it easier for the beekeeper to collect them. In yet another story, tanging convinced the bees to move directly into a hive lined with honey. I have my doubts about that one.

Try it—at worst, you can only feel silly! And it has been seen to work—you never know.

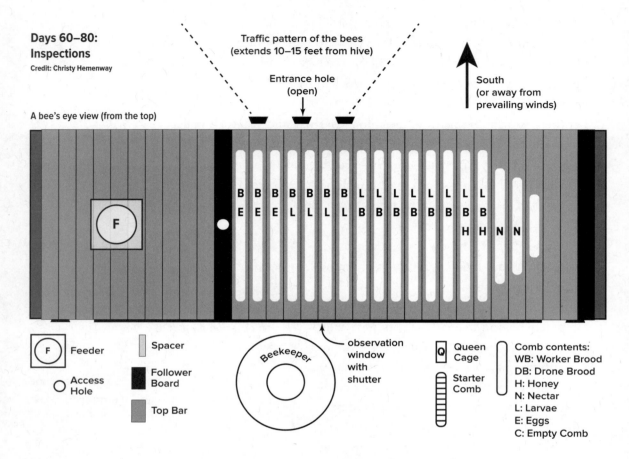

Days 60–80: Inspections
Credit: Christy Hemenway

Traffic pattern of the bees (extends 10–15 feet from hive)

Entrance hole (open)

South (or away from prevailing winds)

A bee's eye view (from the top)

B B B B B B B L L L L L L L
E E E L L L L B B B B B B B
                        H H N N

F

Beekeeper

observation window with shutter

F Feeder

O Access Hole

Spacer

Follower Board

Top Bar

Q Queen Cage

Starter Comb

Comb contents:
WB: Worker Brood
DB: Drone Brood
H: Honey
N: Nectar
L: Larvae
E: Eggs
C: Empty Comb

obvious ability to thrive. Continue to feed if needed, continue to watch for and prevent or correct any cross-combing and be alert for signs of any queen problems or swarm preparations.

Days 70–100: Mid-Season Shift

This mid-season shift is key to managing a center side entrance hive. When you start your bees in the center of the hive, eventually you will need to move the entire hive to one end so that you can utilize all of the space in the hive. But you must still keep the colony building in the initial direction.

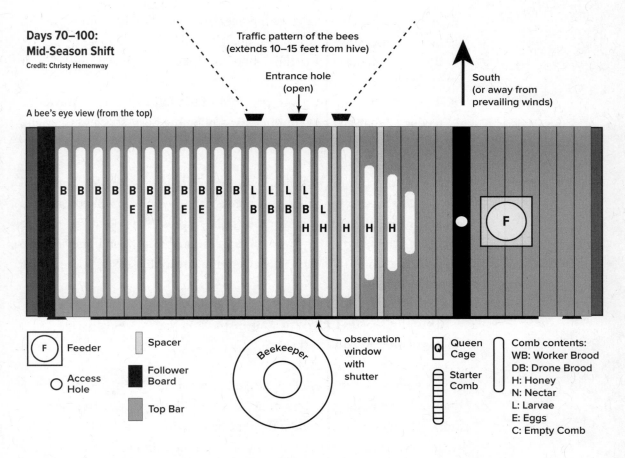

Days 70–100:
Mid-Season Shift

Credit: Christy Hemenway

A bee's eye view (from the top)

Traffic pattern of the bees
(extends 10–15 feet from hive)

Entrance hole
(open)

South
(or away from
prevailing winds)

F Feeder

○ Access Hole

Spacer

Follower Board

Top Bar

Beekeeper

observation window with shutter

Q Queen Cage

Starter Comb

Comb contents:
WB: Worker Brood
DB: Drone Brood
H: Honey
N: Nectar
L: Larvae
E: Eggs
C: Empty Comb

The diagram above shows the hive after you make these changes:

1. Remove the bars to the left that were originally over the feeder.
2. Remove the feeder.
3. Take the solid follower board from the right-hand side, and move it to the far left side of the hive.
4. Move the follower board with the *drive-through window* access hole to the far right-hand side of the hive.
5. Do a bar-by-bar inspection, moving each bar of comb all the way to the left, against the solid follower board, which is now against the end panel of the hive.

6. When you are done, enclose the bee's area with the drive-through window follower board.

7. Put the feeder in place to the right of the drive-through window follower board, if needed.

8. Then put the empty bars to the right of the follower board, above the feeder, so that you can continue expanding the bees' space to the right, as you have been doing.

A mid-season shift inspection moves the entire hive to the left and all of the empty space to the right. This means that the colony will continue to expand in one direction and to put the honey stores to the right. You want to encourage this, as you don't want your bees caught between honey on two sides when it comes time to make a decision late in the season about which direction to move.

> **This mid-season shift is the only inspection that you purposely begin on the brood nest side.**

When your bees begin drawing fatter comb and storing honey, that is the time to begin adding spacers between bars of honeycomb. The bees will draw the honeycomb out wider than the brood comb. The spacers can be inserted between top bars in two ways — set on their thin edge (skinny-wise) or laid down flat (flat-wise), depending on how thick the bees are building the comb (see Chapter 4, page 50).

The addition of spacers will affect the number of top bars that will fit in your hive. When you find yourself needing to remove a bar because the spacers have displaced it and it won't fit any longer, if you have a gable roof, you can leave it on top of the top bars, under the roof, for storage.

Days 80–120: Building Again in One Direction

Continue to march! Add bars into the bees' space, moving the follower board to the right (or to the rear in an end entrance hive) and allowing them to continue to expand the hive. With this much in the way of honeycomb

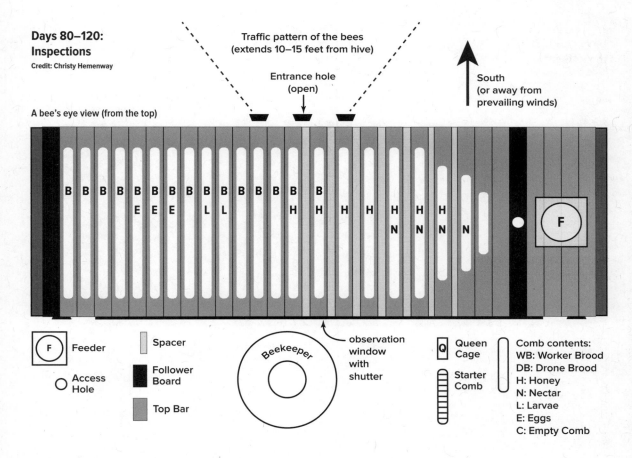

Days 80–120: Inspections

Credit: Christy Hemenway

Traffic pattern of the bees
(extends 10–15 feet from hive)

Entrance hole
(open)

South
(or away from
prevailing winds)

A bee's eye view (from the top)

	Feeder		Spacer		observation window with shutter		Queen Cage		Comb contents:

F: Feeder

Access Hole

Spacer

Follower Board

Top Bar

Beekeeper

observation window with shutter

Q: Queen Cage

Starter Comb

Comb contents:
WB: Worker Brood
DB: Drone Brood
H: Honey
N: Nectar
L: Larvae
E: Eggs
C: Empty Comb

stores in the hive, it is usually safe to stop providing the colony with sugar syrup, but watch carefully what your local weather is doing—during a *nectar dearth* the bees can consume a lot of stores.

Days 110–160: Wall-to-Wall Bees

The hive will soon be very full at the rate that this hive is progressing! This is a very cheery sign—and is a great amount of progress for a first-year top bar hive.

Depending upon when in the season this size has been reached, it may be possible to harvest some honey from this first year hive. If it is still early in the season and there is plenty of forage available and growing season still

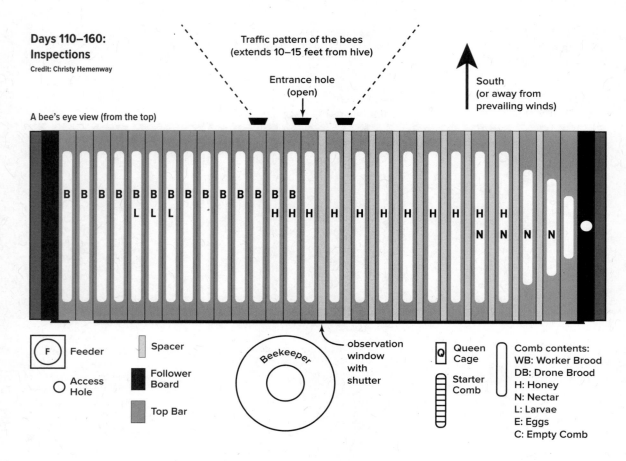

Days 110–160: Inspections
Credit: Christy Hemenway

Traffic pattern of the bees (extends 10–15 feet from hive)

Entrance hole (open)

South (or away from prevailing winds)

A bee's eye view (from the top)

F — Feeder

○ Access Hole

Spacer

Follower Board

Top Bar

Beekeeper

observation window with shutter

Q — Queen Cage

Starter Comb

Comb contents:
WB: Worker Brood
DB: Drone Brood
H: Honey
N: Nectar
L: Larvae
E: Eggs
C: Empty Comb

ahead, it is possible to harvest one or two bars of honey, and the bees will still have time to build it back before the end of the growing season.

It is always best to err on the side of caution. If the bees have as much of their own honey as possible to consume during the winter and they survive the winter, then chances are good that there will still be honey in the hive for you to consume in the spring.

Day 140 through Late Season: Filling the Hive

At this stage of the growing season, the bees prepare for winter. The brood production in the hive will soon be slowing, and so the brood nest portion of the hive will soon be shrinking. As the number of bees that the colony

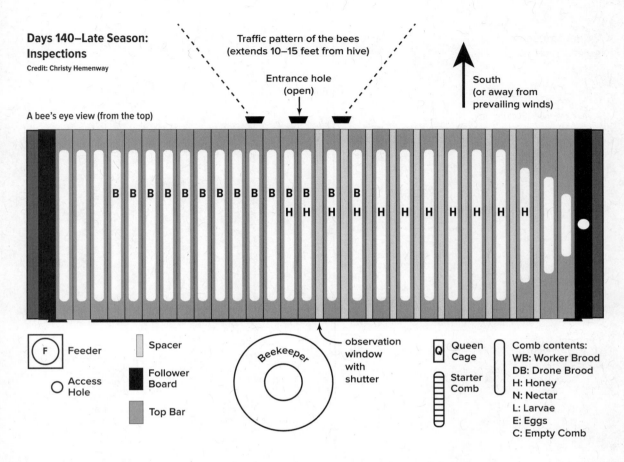

Days 140–Late Season: Inspections
Credit: Christy Hemenway

Traffic pattern of the bees
(extends 10–15 feet from hive)

Entrance hole
(open)

South
(or away from
prevailing winds)

A bee's eye view (from the top)

| | F | Feeder | | Spacer | | Beekeeper | observation window with shutter | | Q | Queen Cage | | Comb contents: WB: Worker Brood DB: Drone Brood H: Honey N: Nectar L: Larvae E: Eggs C: Empty Comb |

F — Feeder
O — Access Hole
— Spacer
— Follower Board
— Top Bar
Beekeeper
observation window with shutter
Q — Queen Cage
— Starter Comb

Comb contents:
WB: Worker Brood
DB: Drone Brood
H: Honey
N: Nectar
L: Larvae
E: Eggs
C: Empty Comb

will be preparing to take through the winter becomes smaller, the bees will begin backfilling some of the brood comb on the right-hand edge of the brood nest with more honey stores .

Winter Shutdown

Prepare for winter shutdown by placing empty combs outside the follower boards. Place the follower boards so that they surround the active colony and their honey stores. Try to ensure that the honey is all to one side of the bees' cluster.

Then prepare the hive for winter, as described next, in Chapter 7.

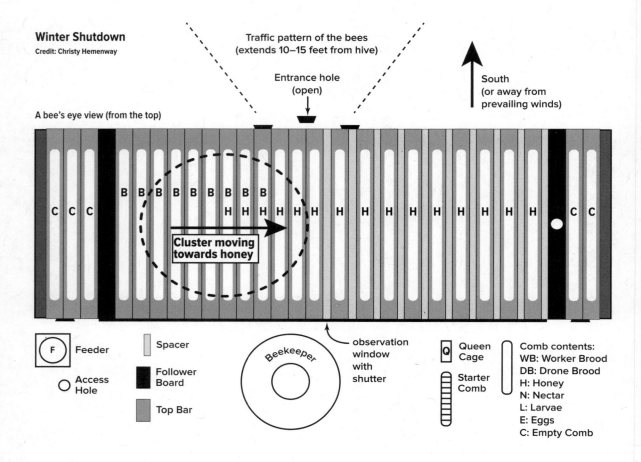

Winter Shutdown
Credit: Christy Hemenway

Traffic pattern of the bees
(extends 10–15 feet from hive)

Entrance hole
(open)

South
(or away from
prevailing winds)

A bee's eye view (from the top)

C C C

B B B B B B B B B
H H H H H H H H H H H H H H H H H H H H C C

→
Cluster moving
towards honey

observation
window
with
shutter

F Feeder

O Access Hole

Spacer

Follower Board

Top Bar

Beekeeper

Q Queen Cage

Starter Comb

Comb contents:
WB: Worker Brood
DB: Drone Brood
H: Honey
N: Nectar
L: Larvae
E: Eggs
C: Empty Comb

CHAPTER 7

Overwintering
Your Top Bar Hive

Winter is a very tough time for honeybees.

The Bee Informed Partnership began studying winter losses in the aftermath of Colony Collapse Disorder, and had this to say:

"Unfortunately, the rate of mortality suffered by overwintering colonies over the last 4 years has been unsustainably high..."[1]

The Partnership also is convinced that:

"...beekeepers, when presented with beekeeper-derived data that objectively show which management practices worked and which did not, will adopt the more successful practices."[2]

When you focus on what is going right and not on what is going wrong—you will get more of what you are focused upon! Which is exactly the right direction to move in.

And if one thing that is going right is chemical-free beekeeping on natural beeswax, then participating in the Annual Winter Loss Survey by the Bee Informed Partnership is a good way for top bar beekeepers in the US to share that information on a national level.

Question #151 on the 2011–2012 Annual Winter Loss Survey included Top Bar Hives as an equipment choice for the first time. This was very

National Winter Loss and Management Surveys

Hive Management

151. What type of equipment do you keep a majority of your hives?

- ○ Standard Langstroth 10 frame hives
- ○ Standard Langstroth 8 frame hives
- ◉ Top bar hives
- ○ Other, please specify

152. Over the production season, how large on average are the brood chambers in your colonies?

- ○ 1 deep (or two mediums)
- ○ 1 deep and 1 medium
- ○ 2 deeps
- ○ More than 2 deeps
- ○ Does not apply
- ○ Other, please specify

Question 151 on the 2011–2012 Annual Winter Loss Survey.

Credit: Bee Informed Partnership. *Winter Loss Survey.* [online]. [cited August 21, 2012]. beeinformed.org /about/tier1/.

heartening and indicates a significant paradigm shift from as few as five years ago. Please make it a point to visit the Bee Informed Partnership website to learn how to be a part of the next Annual Winter Loss Survey.

Why is Winter Such a Challenge?

Honeybees are one of those critters that think ahead — they work during the times when there is plenty of food to store food for the times when there is none. That fortunate trait is why we humans are able to enjoy honey! But when the growing season comes to an end — and there is no more nectar or pollen or propolis available in the world around them — the bees must then rely on what they've been able to store up inside the hive to get them through the winter.

Whether their stores prove to be adequate depends upon many variables — the weather, the success of the growing season, the length and severity of the winter, the amount of honey harvested by the beekeeper, the amount of food left in the hive, the size of the colony — and plenty of other things we're not completely aware of.

Hives in areas where winters are harsh will benefit from having additional protection from the winter itself. The need for this protection and the intensity of it of will vary based on the geographical location of the hive.

When to Start Preparing for Winter

Winter preparation actually begins when the bees are installed. Planning ahead during the initial setup of the hive, managing the building of comb through the season and being aware of the logistics inside the hive during cold weather will all add up to being prepared when suddenly, fall arrives and you must endure the next several months without your bees.

Unidirectional Bees

A primary concern in a horizontal hive is to manage the hive so that the bees consistently build in one direction and locate their honey stores on one end of the nest—opposite, but next to the the brood nest. In Chapter 6, the figures on pages 108 and 109 show how to specifically manage your hive for this result as the colony grows.

There are two reasons for this concern:

Bees can't cross empty space in the cold. When bees are clustered during the winter, they are unable to move as a group across empty space in order to get to a source of food on the other side of the hive. As winter approaches, it's best if bees are right up against their food source—and that food source should be unbroken and all in one direction. There are many sad stories of hives starving to death with food only inches away. You can prevent this by shepherding the building of the colony as it progresses through the season.

Bees are not quick decision makers—they operate as a *hive mind*. That's why it's best in general, but especially in winter, to avoid having honey on both sides of the brood nest in a horizontally oriented hive. It can take your bees some time to come to a consensus about moving and about which direction to move—and that time can endanger their survival in winter. In the words of Kim Flottum, "This is why bees don't rule the world!" So helping to ensure that their food supply is in only one direction is a helpful thing.

Wind is a Four-Letter Word

Protecting your hive from the wind is probably the single most important thing you can do as a beekeeper to support your bees during the winter months. A constant winter wind can be very desiccating and draining. There are probably as many ways of protecting your top bar hive from the wind as there are beekeepers. Some of the best and most effective ways that I have seen of providing wind protection include the following:

Natural Planting

A stand-alone hedge, or a row or a circle of bushy, branching trees provides a good windbreak. This doesn't have to be completely windproof—but it needs to be thick enough to break the wind and keep it from being a constant drain.

This successfully overwintered hive was surrounded by a natural windbreak.
Credit: Christy Hemenway.

Privacy Fence

A hive tucked into the corner of a backyard surrounded by a privacy fence that is taller than the hive is well protected from wind. This is very successful in suburban areas.

Hay Bale Fort

Stacking bales of hay or straw around the hive has proven to be tremendously successful. Place the bales close to the hive, but not so close that the entrance is completely blocked. Be particularly sure to install a mouse guard in this instance, as mice are also likely to be attracted to the shelter of the bales.

This successfully overwintered hive was tucked into a corner of a fenced backyard.
Credit: Pat Beckett.

This temporary surround can be constructed out of mulch hay or straw.
Credit: Kimberly Strader-Marshall.

Tarp Skirt

I've received good reports of success with this method of protection as well. Attach the upper edge of the tarp to the upper edge of the hive body, especially on the windward side, and either weigh it down or attach it to the legs near the bottom. Leave the entrance accessible.

Foamboard Insulation Panels

Foamboard panels, cut to fit the sides and ends of the hive and held in place with bungee cords or screwed directly into the wood of the hive, do a great job of breaking the wind and insulating the hive from low winter temperatures. Be sure to leave the entrance hole(s) uncovered.

Tar Paper Wrap

This is a technique borrowed from Langstroth wintering practices. If the hive is wrapped completely, the tar paper serves to act as a reasonably good windbreak. The tar paper can be wrapped around the body of the hive and held in place with staples. It helps to combine this with another windbreak, such as a hedgerow or bush.

Food Is a Four-Letter Word Too

As temperatures decrease and the growing season draws to a close, the bees need to have "plenty" of honey stores. That's a pretty open-ended term, I know. There is endless speculation and debate among beekeepers concerning how much honey is required to get a hive through winter. This is an elusive number because it depends on so many

A tarp skirt is another good way of protecting your hive from wind.
Credit: Bill Price.

Foam insulation cut to fit and attached to the hive offers good protection.
Credit: Tom Kruzshak.

Tar paper wrapped round the hive body helps to break the wind.
Credit: Christy Hemenway.

variables: the length of the winter, the size of the colony, how cold it gets, how windy it is.

In a first-year top bar hive, if your bees do not completely fill a 30-bar hive body, then I strongly recommend that you leave the bees all of their honey to eat over the winter. If they did fill the hive completely with comb — the condition I like to call *wall-to-wall bees* — then you can cautiously harvest individual bars in mid to late summer, keeping a careful eye on the weather and the nectar flow and being careful not to harvest more than the bees will likely be able to replace before the first frost or the end of the local growing season.

One of the driving methodologies behind healthy and natural top bar hive management is to avoid feeding sugar syrup and to let the bees eat their own honey whenever possible. After all, where is the long-term sustainable benefit of taking the bees' healthy, nutritious, natural food and then having to feed them junk food in order to keep them alive (D'oh)?

It is commonly said that a top bar hive will overwinter on less honey than is required by a Langstroth hive, and in practice that is what I have seen as well. But again, there are so many variables involved that it is impossible to offer specific numbers to answer the question of how much honey that is.

In spite of all these variables I know that thinking beekeepers still want some sort of benchmark. So let's say, as a rule of thumb, that the hive should have a minimum of six to eight full bars of honeycomb going into winter. This is in addition to any honey that the bees have backfilled into brood comb as the colony reduced its size when the growing season came to a close and brood rearing wound down. This many bars should be enough to manage a fairly serious winter (says the author from Maine).

If your hive does not have at least six full bars of natural honey stored, then you should feed, starting in late summer or early fall. There are several food sources to consider when feeding your bees in the fall, going into winter.

Reserved Honey

If you reserved combs of honey made by your bees earlier in the season, you can give these back to them now. It's a good idea to store these combs in the freezer to protect them from wax moths and other pests. One logistical complication with this plan is that each bar of honey that you reserve means you will have removed a top bar from your hive — so plan accordingly.

If temperatures allow and you have liquid honey available that was harvested from your own bees, or from a source that you are confident was not treated with antibiotics (see Chapter 5), you can feed this to the hive as well, much like sugar syrup. It may require that you dilute the honey somewhat to get it to flow inside the feeder.

Liquid honey can be fed inside the space in the hive that the bees are occupying, provided some combs are less than the full depth of the hive. Use a shallow dish with a *bee ladder* (made of a piece of hardware cloth bent into a checkmark or sliding board shape) to prevent the bees from falling into the honey, where they will get stuck. This ladder-and-dish arrangement gets the food closer to the cluster and makes it more accessible in cooler temps.

Sugar Syrup

If temperatures are warm enough, the bees will consume sugar syrup. But in the fall, you want to be certain that you increase the ratio of sugar to water from the spring formula of 1:1 to the fall formula of 2:1.

2:1 FALL SUGAR SYRUP RECIPE
1. Boil ten cups of water.
2. Remove from heat.
3. Add ten pounds of granulated sugar. Stir to dissolve.
4. A dash of lemon juice can also be added as a preservative or herb tea added as a medicinal element (see Chapter 5).

This simple syrup can be stored in the refrigerator for up to several weeks. Bring it to room temperature before putting the syrup in the hive.

Dry Sugar

Bees can consume dry sugar during cool fall temperatures, and will sometimes do that more readily than they will consume syrup. The closer the sugar is to the location of the bees' cluster, the more likely it is that the bees will be able to use it. If necessary, you can put dry sugar directly on the floor of the hive in the cavity the bees are occupying. It's a good idea to lightly spray the sugar with water to cause it to crystalize a bit and develop a light crust, otherwise the bees may simply carry the dry sugar out of the hive as part of their housecleaning tasks.

Fondant or Bee Candy

Fondant is a way of feeding sugar in a semi-solid form. Fondant does not introduce the amount of moisture into the hive that syrup does, which is a benefit late in the season. The fondant can be placed on the bottom of the hive if the bees have not yet clustered due to cold temperatures—or it can be attached to a follower board, or suspended from a top bar, located as close to the cluster as you can get it—touching if possible.

Fondant can be purchased at cake supply stores, and sometimes at craft stores as well. Avoid fondant that has color added.

You can also make your own fondant—this recipe is reprinted here with permission from and thanks to John and Ruth Seaborn of Wolf Creek Apiaries:[3]

Fondant Bee Candy Recipe

Use pure cane sugar ONLY in this recipe—not raw sugar or turbinado sugar, or brown sugar, as these contains solids that are bad for the bees' digestive system.

1. Use 1 part water to 4 parts sugar.
2. Add ¼ tsp. vinegar per pound of sugar. (1 pound of sugar = 2¼ cups). The vinegar helps to break down the sugar as it cooks and will be evaporated. Add ¼ tsp. of Real Salt (sea salt with 60 naturally occurring minerals) per batch.
3. Bring to a boil, stirring constantly until boiling begins. Making the candy without stirring will yield a transparent gel that will be extremely sticky.
4. Boil covered for 3 minutes without stirring.
5. Boil until mixture reaches 234°F. Going over this temperature will cause the mixture to caramelize and will be harmful to the bees.
6. Remove mixture from heat and cool to 200°F.
7. Quickly whip with a whisk until whiteness occurs. If preferred, ½ cup of Ultra Bee Pollen Substitute may be added at this time.
8. Quickly pour onto waxed paper having a towel beneath. Be sure that the towel is flat, not fluffy, since this will lessen the cake's width. This method will make a nice cake.
9. Allow the fondant to cool undisturbed.
10. Remove waxed paper and store each cake in a plastic bag in the refrigerator. The cakes can be handled as plates, but may be a little fudgy. They will be completely white with whiter areas inside. Tiny crystals will shine from a broken edge.

This recipe—made with 1 cup of water to 4 cups of sugar—will yield a cake of fondant about the size of a large dinner plate, approximately half an inch thick.

To feed fondant in a top bar hive—the cake of fondant can be attached to a follower board using mesh, or rubber bands. It can also be suspended from a top bar inside a manila folder or other similar "envelope" with slits cut into the sides so the bees can access it.

Thanks to John and Ruth Seaborn at Wolf Creek Bees for sharing this fondant recipe with us.

Cold Is a Four-Letter Word as well

Containing the Colony

The follower board(s) should contain the colony's brood nest and honey stores as they prepare for winter, but as the season comes to a close, you may find that your bees have made some empty comb at the edge of their space. This comb can be moved to the outside of the follower boards surrounding the nest and held there in reserve for next year. Storing it outdoors in cold temperatures helps to prevent issues with pests. (See Chapter 4, page 55, and Chapter 6, pages 108 and 109.)

Managing the empty comb in this way helps to make the cavity that your bees must heat through the winter just that much smaller; this is helpful for the overwintering cluster.

Other than this extra comb being moved to the outside of the colony's overwintering space, I advise against shuffling combs around within the bees' nest late in the season. The configuration they have created with the individual combs had a special purpose according to the bees' logic, and it seems wise not to tinker with their natural processes.

Solid Lid of Top Bars

The bars in a top bar hive all need to touch each other, and they need to cover the entire space occupied by the bees—the space between the follower boards, or between a follower board and the actual end of the hive. There should be no gaps between the bars. The bees should be able to occupy, and have to heat, only the space beneath the top bars.

Adjustable Bottom Board

On hives with a removable or adjustable bottom board, you should raise or otherwise close this board. If the gap between the hive body and the bottom board is significant, this gap can be weather-stripped or sealed shut with duct tape to reduce drafts.

Inside the Roof

If your hive has a gable or peaked roof creating a space above the bars inside the roof, this space should be filled in order to prevent air movement above the top bars. This can be done with any number of methods—the one I have found simple has been to take a piece of insulation and enclose it in a plastic bag. Tuck this bag inside the roof to fill the space.

Outside the Roof

A good habit to develop as a matter of course is to strap down the roof of the hive. This will keep a strong, sudden, winter wind from blowing it off. I strongly recommend this even though the roof can weigh 20 pounds and fit so closely that one would not think it necessary. It only takes one instance of finding a hive missing its roof to make a beekeeper wish that she had taken this one simple step. I use a long nylon strap with a metal spring clip. The strap can be run around the hive body itself, which also serves to tighten an adjustable bottom board, or around the hive and down to an anchor below.

Entrances

As the first frost threatens and daytime temperatures drop into the 50°F range, all but one of the entrances should be closed. If your hive has multiple round entrances, these can easily be closed with corks. If your hive has a long slot entrance, a piece of wood with a cutout in it, sized to fit, should be inserted into the slot, leaving only a limited place for the bees to enter and exit.

One entrance must always be left open because the bees will need to be able to fly in the event that the temperature rises above 50°F on a sunny winter day. This will be the sort of day when bees make a *cleansing flight*. The bees do their best never to defecate in the hive, and so the opportunity to make such a flight is important.

You will never regret securing the roof of your hive.
Credit: Christy Hemenway.

Mice

Mice love to find an unprotected beehive in winter; it makes a perfect home for them. For one thing, the space is sheltered and enclosed. For another,

Bees and Laundry

The kind of day that bees like to take a cleansing flight is often the kind of day that winter-weary folks like to hang laundry on the clothesline. The day will be bright, breezy, warmish...a relief from sharp cold temperatures and snow. Be aware, though, that if your fresh, clean laundry is in their flight path, bees taking a cleansing flight are likely to leave many small yellowish-brown droplets of bee poop on it.

One-quarter-inch hardware cloth works well as a mouse guard over a single entrance.
Credit: Christy Hemenway.

with the bees clustered tightly on the combs in cold temperatures, the rest of the contents of the hive are fair game as a source for food. It's very disappointing to open a hive in the spring and find the remains of all the bees' hard work in pieces on the floor of the hive and a fuzzy nest full of fat, warm, happy mice in a corner of the hive.

To prevent mice from getting into a hive with round entrances, use a piece of ¼-inch hardware cloth, cut to fit and stapled over the open entrance, as a mouse guard. This size is large enough for the bees to enter and exit the hive, but too small for mice to be able to enter. The best way I have found to install this mouse guard is first to place a piece of tape over the area on the hardware cloth that will be covering the hive entrance. This prevents the bees from exiting the hive while you are stapling the guard in place. The impact of the stapler will often bring defensive guard bees out to investigate even if the temperatures are low. After you've attached the hardware cloth, and done any other winter prep work necessary, remember to pull off the tape before leaving.

When to Stop Preparing for Winter

Preserving the Propolis Seal

When should beekeepers stop inspecting for the season? It is important to inspect the hive near the end of the season in order to check on the arrangement of the colony and the amount of stores available. However, it's also important to realize that as winter approaches the bees are striving to seal up gaps in the hive with propolis as part of their preparation for the winter—and this includes the joints between the top bars.

As the season progresses, you will notice a difference in the quality of the propolis during inspections. In the spring and early summer, the propolis is soft and gooey—almost like pizza cheese. When you put the bars back together after inspecting a hive in the summer, the propolis can be so warm and sticky that it joins things right back together as you inspect. Later in the season, the propolis becomes much more brittle, so that when you insert your hive tool between top bars and give the twist required to separate them, there is an audible pop or snap as the propolis breaks. In the fall, the bees are no longer able to gather the tree resin necessary to replace the propolis, so the beekeeper should not destroy this important seal.

In Maine in recent years, this brittle popping stage has been occurring in mid-October. It is better not to disturb the bees after this time, other than to add fondant if necessary. You are likely to find, too, that the bees get pretty aggravated by being disturbed at this point in the beekeeping season. Best to stop when the propolis pops!

Sitting On Your Hands

Once you've winterized your hive—done the best you could to bolster them with some food stores if they were short and provided protection from mice and from the wind—there is little else you can do. In fact, opening the hive can do more harm than good. So now is the time to do the thing that is toughest for beekeepers to do: That is to sit on your hands and not mess with your bees until spring comes!

When Spring Finally Comes

Temperatures above 48°F should show you at least a few bees flying. Check to be sure that the entrances secured by mouse guards are not blocked by dead bees.

Look for activity, but don't be too quick to judge. Even a peek through the observation window may not show you any bees. Bees can cluster very tightly together and be very difficult to spot.

On a still, sunny day above 50°F, you can sneak a quick peek into one end of the hive to learn what you can about the remaining food supply.

Since you don't want to go deeply into the hive at these temperatures, breaking apart the propolis and exposing the earliest beginnings of brood rearing, be very conservative at this stage. If you think the bees need food — provide some. At the temperatures typical of this season, fondant is the best idea. A fondant feeder built into a follower board can help you to get food as close to the cluster of bees as possible.

When daytime temperatures reach 50°F consistently, and begin to allow for daily flight, and when there is forage available and you are convinced that spring has truly arrived — then you can reverse the winterization steps you took last fall:

- unwrap or uncover the hive
- move hay bales to allow access
- remove any insulation from inside the roof
- remove mouse guards

When nighttime temperatures get consistently into the 50s°F
- open additional entrances, as activity requires
- lower an adjustable bottom board, if your hive has one

Stay alert for abrupt changes in the weather. A sudden serious drop in temperature can be a disaster for bees that have begun to raise brood — especially if food supplies are short inside the hive or not located close to the cluster.

April

April is often said to be the cruelest month in beekeeping. A hive that appeared to have overwintered and was beginning to thrive can sometimes lose its balance in April and be gone before May. So in April be optimistic, but don't be completely convinced until you're sure your hive has turned the corner. It is often warm enough to feed syrup in April if needed, and it may be best to err on the side of caution in this treacherous month.

A successfully overwintered top bar hive can build up enthusiastically in the spring and be ready to swarm in May. Don't underestimate the bees as the growing season begins again. Trees are some of the earliest sources of pollen along with the humble, ubiquitous dandelion. Get familiar with the seasonal bloom where you are keeping bees. This way you can track what food sources your bees have, and this will help you to estimate whether you are safely over the hump between winter and spring.

An overwintered Gold Star hive, bringing in loads of spring pollen.
Credit: Christy Hemenway.

Success!

By May, if your bees have overwintered and your local growing season is now back in full swing, you can now be justifiably proud of having shepherded a top bar hive into its second year. I collect overwintering success stories — if you'd like to share yours with me, you can write to me.[4]

Congratulations!

CHAPTER 8

Treasures of the Hive

In addition to the magic of the bees themselves, top bar hives provide two additional treasures from the bees' natural systems — honey and beeswax. Digging deeper, another treasure can be found in the venom of the bee's sting, designed to help defend that honey.

Ah, Honey!

Just what is honey? Only the most amazing natural sweetener on earth, don'tcha know? Honey is that sweet golden liquid that honeybees make and store during times when there is plenty of forage — in order to get themselves through times when there is no forage. The making of honey, sometimes jokingly described as "bee vomit," begins with bees foraging among flowers to gather nectar.

It's been said that the average worker honeybee gathers only enough nectar in her entire lifetime to make $1/12$th of a teaspoon of honey. So needless to say, it requires an amazing number of trips from the hive to the flowers and then back to the hive to collect enough nectar to create one cell of honey.

Without getting too deep into the science of it, it works something like this: Plants need pollination, but they can't move — so they need to attract bees. To attract the bees, the plants create nectar. The bees gather nectar from the flowers. They add enzymes to the nectar. They store the nectar in open hexagonal cells in their beeswax combs. They evaporate the moisture from the nectar-enzyme combination by fanning — moving massive quantities of air through the hive — and reducing the moisture level to approximately 18%. Then they *cap* each cell by building a wax lid over the cell, sealing in the reduced liquid, which has become honey. The honey is stored in honeycomb until the bees need it for food, or the beekeeper or some other honey-loving critter harvests it.

Meanwhile all this nectar gathering for honey production helps to pollinate the plants, which will then be able to set fruit, create seeds and reproduce. Yep, amazing is a good word for honey!

Beekeeper Focus

Beekeepers who focus on producing honey, especially as a source of revenue, work with their bees in ways that are designed to maximize the production of honey. Top bar beekeepers tend to focus on the health of the bees and the natural systems at work in the hive, and place less emphasis on getting honey from their hives.

The fact is honeybees make honey. And they can often make enough of it that the beekeeper can remove surplus honey from the hive without endangering the bees. Learning to strike that balance is crucial—since it doesn't make good sense to break the natural system at work inside the hive in order to harvest too much honey. That's like killing the goose that laid the golden eggs.

So here are some tips on harvesting this golden treasure from your top bar hive.

When and How Much to Harvest?

When I founded Gold Star Honeybees®, I made a point of saying, "It's not about the honey, Honey—it's about the Bees!" I even had t-shirts printed that said that. This was my way of saying that if you have healthy honeybees, you'll get honey—but that the primary focus of beekeepers needs to be on healthy honeybees.

Successfully overwintering a hive on the bees' own healthful, natural honey is surely a primary goal of natural beekeeping—and so I always recommend that no honey should be harvested from first-year hives. Leaving your hive with all of the honey the bees are able to make in their first season

A brand new comb of honey being capped by the bees.
Credit: Christy Hemenway.

is a sound practice. This is based on the expectation that a first-year hive, started from a package, doesn't generally fill the entire hive with comb, so usually there is no surplus honey.

Occasionally a hive goes wall to wall with comb in one season, and there is surplus honey that you can harvest during its first year. Then it becomes a matter of gauging whether the bees will be able to replace the honey that is harvested in the time remaining before winter arrives.

One needs to develop a bit of intuition about this—a combination of your local weather, knowing what is available for forage where you live and when—and being aware of the state of things inside your hive. All these things will lead you to an intelligent decision. While you develop your intuition and knowledge, it helps to be conservative.

In a top bar hive, the beekeeper selects individual bars of honey for harvesting. It is important that the bars of honeycomb chosen for harvest be capped—in other words, that the cells have been sealed with the wax lid made by the bees. This indicates that the honey is ripe. Ripe honey has an amazing reputation for never going bad or spoiling.

Honey that is not yet ripe (in other words is uncapped, or still in the nectar stage) contains more moisture than ripe honey, and it can ferment. And while that doesn't really ruin it, it's different from ripe honey, the taste may be quite strong and the texture will be very runny. Best to leave any unripe honey for the bees to finish.

How to Harvest Honey

Since the honeycomb is removed from the hive and destroyed in the process, harvesting from a top bar hive becomes, *ipso facto*, a dual harvest of both honey and wax. To harvest, the capped honeycomb is removed from the hive, the bees are brushed from the comb and the comb is cut from the bar. The bar is replaced in the hive for the bees to rebuild on. Be sure to say thank you to your bees!

Then, one of two things can be done with the harvested comb.

Cut Comb Honey

The full honeycomb can be cut up and stored just as it is, in a covered container. This is known as cut comb honey or honeycomb. I like to call it "bees on toast," a phrase I borrow from Phil Chandler, my original top bar hive mentor. If you've ever had warm toast with honeycomb spread on it…well, there are few things in the world that you can have for breakfast that are as magical or amazing as that.

Bees on toast—a special treat.
Credit: Christy Hemenway.

Liquid Honey

Honeycomb can be processed into liquid honey via the *crush and strain* method. The honey harvest kit described on the opposite page utilizes two five-gallon buckets.

Bottling Honey

The best honey jars are made of glass, have wide mouths and screw-top lids. Unprocessed, unheated honey becomes quite thick in the jar, and it just won't pour like store-bought liquid honey

- Bucket #1—This is the top bucket. Drill a series of ½-inch holes in the bottom of this bucket, within a four-inch circle in the center of the bucket's bottom.
- Bucket #2—This is the bottom or collector bucket. Drill a four-inch hole in center of the lid of this bucket. (It's really convenient if this bottom collector bucket is also equipped with a valve near the bottom known as a *honey gate*. This makes it much easier to bottle the liquid honey, but without a honey gate you can use a ladle or spoon).

1. Put the drilled lid on the bottom bucket.
2. Set the top bucket on top of the drilled lid of the bottom bucket.
3. Line the top bucket with a fine nylon strainer. These are available from the hardware store as paint strainers—they come in a five-gallon size, with an elastic edge at the top. This strainer will be fine enough to catch wax and bee parts, but it will let the honey and the pollen through.
4. Put the honeycomb into the top bucket. Use your hands or a blunt tool to crush the comb. A potato masher is a common choice; a wooden spoon works as well. Mash the comb thoroughly. Break open all the wax cells. This releases the honey. The honey will then drain, at honey's fairly leisurely pace, from the top bucket through the strainer and into the bottom bucket, where it will collect and can be bottled.

A crush and strain honey harvest kit— very low-tech!
Credit: Brian Fitzgerald.

5. When you're done crushing, put the lid on the top bucket and leave it be for a day or so. The honey will drain more quickly if you do the crushing and straining in a room that is comfortably warm. Then, provided your gravity bill is paid up, when you come back you will have honey in the bottom bucket and wax in the top bucket. Voilà—dual harvest!

Bottling the natural finished product—liquid honey.
Credit: Adam McNally.

that's been pasteurized and overheated, so you'll want to be able to reach deep into the jar with a spoon.

Honey can (and usually does) crystalize over time. The best way to turn it back into liquid is to immerse it in a warm water bath. This is one reason that storing honey in a glass container is a best practice—glass can take the heat. Please don't microwave the honey to liquefy it. Be patient, and use a gentle warm heat.

I recommend against using plastic bottles for honey—partly because glass is a cleaner, more benign material, but also because of the risk of introducing Bisphenol A into your very clean, natural honey! Especially, please don't microwave a plastic bottle of honey—for the honey's sake yes, but also because of the effect of the microwave on the plastic bottle, and then the effect of the plastic bottle on the honey.

Healthy and Nutritious—It's the Real Deal

Best of all, honey that's been harvested in these ways will retain all of its wonderful nutritious and medicinal properties, because it has not been heated to the point of being pasteurized, nor has it been filtered so intensely that the pollen has been removed from it.

Save it for Bee Food

Honey—their own honey—is the best food available for your bees. The honey can be harvested as described above and then fed back to the bees using the same feeder used to feed sugar syrup. It may require the addition of water to make it liquid enough to flow, but honey is the best food for your bees.

If you have extra top bars available for your hive, so that you are able to replace any bars that you remove, then honeycomb can be stored and later given back to the bees, in the comb, still on the bar.

AFB WARNING!

You may use honey produced by other bees to feed your bees—however, be aware of this serious caveat. You must be very certain of the source of the honey, and more specifically, you should be aware of any medications that were used in the hive where that honey came from. The reason for this strong warning is that honey that comes from hives that have been treated with antibiotics to prevent American Foulbrood (AFB) may not have shown the symptoms of the disease, but the honey made by these bees may still contain the spores that carry the disease. The antibiotics will have suppressed the germination of the spores, but the spores can still be present.

In your hives, however—assuming you choose not to treat prophylactically with antibiotics—these spores will be able to germinate, and this can infect your hive with American Foulbrood. Since the typical treatment for American Foulbrood in most locales is to destroy the hive—usually by burning—this is definitely something you want to avoid. If you cannot be certain of the source of the honey and confident that it has not been treated with antibiotics, it might be considered a better course of action to rely upon sugar syrup until your bees are able to make their own healthy honey.

Amazing Remedies from Honey and Venom

Pollen Allergies

Honey from the same plants that cause your allergic reaction to pollen, honey that has been harvested without being heated or over filtered — thus retaining that pollen — can be very effective medicine. It can help to relieve symptoms such as sneezing, runny nose and itchy eyes — without the annoying side effects of allergy medicines.

I can personally attest to honey's effectiveness on allergies: After decades of allergy shots and over-the-counter medications, my lifestyle now includes my own locally raised honey. With a daily spoonful of honey in my breakfast tea, honeycomb spread on warm toast, used as a sweetener in baking and cooking and the occasional yummy spoonful straight out of the jar — I have gone from being nearly incapacitated by my pollen allergy symptoms for weeks on end to being only mildly aware of them for about a week of the spring or early summer.

It's very important that the honey you use with allergy control in mind be from bees in your local stomping grounds since you're allergic to pollen in your local area. This gives new meaning to the thought of "buying local," doesn't it? Honey from other places may be incredibly tasty, or may, like manuka honey made from the flower of the tea tree, have powerful medicinal properties. But when it comes to allergies, unless the honey contains pollen from the plants you are allergic to, its efficacy as an allergy remedy is nil.

Wounds and Burns

The next time you find yourself with a minor cut, scrape or burn — before you reach for a synthetic antiseptic ointment, reach into your honey jar and put a bit of honey on the wound. The healing properties of honey will reveal themselves rapidly: Wounds will heal quickly and cleanly. Honey, when it comes into contact with the sodium and the higher pH level of human skin, begins to release hydrogen peroxide, making it antiseptic,[1] and it has been seen to be effective as a natural antibiotic, with no known side effects. Honey is wonderful topical medicine.

Apitherapy—Bee Venom Therapy (BVT)

While modern medicine takes no official stand on the efficacy of bee venom as a treatment, the list of anecdotal evidence and testimonials from people who have benefited from it is a long one. Among the issues that have been helped using apitherapy are arthritis, Lyme disease, multiple sclerosis, various tendon and muscle injuries, skin problems and pain. The best reference I have found for this amazing therapy is the American Apitherapy Society's website.[2]

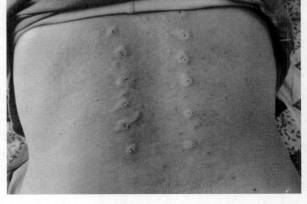

Apitherapy can include applying deliberate bee stings to inject the venom of the honeybee into the body.
Credit: Phillip Dalto.

Rendering the Beeswax

By the time you've kept bees in a top bar hive for a year, you've likely collected some bits of broken comb; maybe you've harvested some honey and have the crushed wax bits that remain from a crush-and-strain extraction and you've probably been saving this beeswax. By now you may even have a sizable container of it waiting for you to do something with it.

Here's what I suggest:

Gather together some tools that you will dedicate solely to the rendering of beeswax. These items won't be useful for anything but rendering beeswax in the future—so be sure not to sneak anything out of the kitchen that your spouse is likely to miss—because it will never be the same again.

- a large metal pot—stainless steel is best
- a large, fairly coarse strainer with a handle—if it just fits into a five-gallon bucket, great
- a plastic shower curtain or tarp to protect the floor if you are working indoors
- a propane-powered two-burner camp stove if you are working outdoors
- a stir stick of some kind—a wooden spoon can work
- a couple of five-gallon buckets

Rendering beeswax can be quite messy—it seems to get everywhere despite your best attempts to contain it. So in the interests of peace in your household and not having to spend a full day cleaning your kitchen, it's wise to work outside if at all possible. A two-burner propane-fueled camp stove does a good job of heating the water that you will use to render the wax. If the flame of the camp stove is affected by wind, a windbreak of concrete blocks or some similar non-flammable item is easily arranged.

If you are working inside on your kitchen stove, protecting your floor is an important first step. The heat of the wax can adversely affect vinyl tile, and other surfaces, such as ceramic tile or wood, will require scraping to remove any spilled wax once it's cooled. This scraping can scar surfaces. A plastic shower curtain or a tarp laid down in your working area will provide protection. Be careful not to trip over any of the edges of the tarp. In the event that you do need to scrape wax, a thumbnail, credit card or blunt-edged putty knife can be helpful in getting it up.

> **Warning! Beeswax is extremely flammable; it should never be heated by itself in a pot directly over a heat source. Always use water. Since wax floats, the use of boiling water to melt it simplifies the rendering process considerably.**

Begin the rendering process by taking the large pot and filling it approximately one-third full of water. Put it on the heat and bring it to a boil. As the water nears the boiling point, you can begin adding comb or crushed wax. Avoid letting any large clumps of sticky, crushed wax fall into the boiling water as it may splash back at you, causing a serious burn. You can continue to add crushed wax and comb until the level of the contents in your pot approach about two-thirds of the pot's capacity. Do not add more than this, as the wax and debris may roil and tumble while boiling and may cause the contents of the pot to splash over the edge.

If you are rendering brood comb, you will find that there will be a significant amount of *slum gum* (brown debris) rising to the surface of the water as the wax melts into the boiling water. These are the cocoons that the bee pupae spun around themselves as they began their metamorphosis, along with other debris that winds up in the wax. You will see this slum gum rise to the surface and tumble there as the wax boils. Use your wooden stirrer to break up these lumps and to keep the contents moving.

Boil long enough to be certain that all of the wax has completely melted: 20 to 30 minutes is generally adequate.

Near where the wax is boiling, set your coarse kitchen strainer into the five-gallon bucket. Be sure to place these on the tarp if you are working inside. It's best if your kitchen strainer is large enough to nearly fill the entire opening of the bucket, as this will help to prevent the strainer from falling into the bucket as you begin to pour the wax and water into the strainer.

When you are ready to pour the wax through the strainer into the bucket, use either old potholders or old towels to protect your hands. Lift the pot of wax and water and pour it carefully, but reasonably quickly, through the strainer and into the bucket. If you are rendering brood comb, the strainer will fill with cocoons fairly quickly. These can be stirred, squeezed and agitated to encourage the wax on its way through. When the contents of the pot have all been poured into the bucket, press your wooden stir stick or other tool against the strainer to express as much of the liquid wax as possible.

Be careful not to step in any spilled or splashed beeswax. It readily adheres to the bottom of your shoe and will get tracked throughout the house.

Carefully heat the wax pot, melting the honeycomb in the water.
Credit: Christy Hemenway.

Strain the wax and water through a coarse kitchen strainer into a five-gallon bucket.
Credit: Christy Hemenway.

The finished product—a disk of pure, untreated beeswax.
Credit: Christy Hemenway.

It is especially difficult to remove beeswax from carpeting—so avoid this if at all possible.

Let the bucket of wax and water cool. Wax floats in water. You will see the wax, as it cools, rising to the surface and taking on the characteristic yellow color of beeswax. When the wax and water have both cooled completely and the wax has hardened, you can release the wax from the sides of the bucket by running a thin-bladed knife around the circumference of the bucket. Then remove the disk of wax by pushing down on one side of it, grasping the other edge and lifting it out.

The disk of wax will have a layer of fine debris on the bottom, which can be brushed off and discarded in the trash. The water in the bucket will be quite dark brown. The slum gum contents of the strainer will still contain residual wax and can be saved and used as an excellent fire starter.

After this first rendering, you may want your wax to be even cleaner. Carefully scrape away as much of the debris from the bottom as possible. If there was a great deal of debris on the bottom of the cooled wax, you may need to render it again. Use the same process as you did for the initial rendering, but this time, in preparation for straining, lay several layers of cheesecloth into the strainer. Be sure to crisscross them so that the fibers alternate, making a finer mesh. Carefully melt the wax into the boiling water, then pour through the cheesecloth-lined strainer and again allow it to cool.

This beautifully rendered beeswax can be used to make beeswax candles or added to salves, ointments, other herbal preparations, and in encaustic or hot wax painting, among other uses.

Candles

Beeswax candles are different from paraffin candles, which are essentially petroleum products. Beeswax candles burn longer, without soot and without emitting unpleasant and toxic fumes into the air. Beeswax candles are actually said to purify the air around them when they are lit.

Salves and Creams

To make a soothing herbal salve, mix the following together in a small pan:

Sweet almond oil — 3 oz.

Jojoba oil — 1 oz.

Canola oil — ½ oz.

1. Melt up to 2 oz. of clean, untreated beeswax into this mix to make a base.
2. Allow the base to cool, and check the consistency of the salve. To thin, add more sweet almond oil and reheat to blend. To thicken, add more beeswax and reheat to blend.
3. As the mixture is cooling the final time, add 40 drops of an essential oil of your choice.

Lavender is very soothing. Other essentail oils have healthful properties as well.

Household Uses

Other uses for beeswax include the waxing of thread when sewing by hand; this serves to strengthen and smooth the thread. Beeswax also helps to ease the movement of a stuck drawer.

These treasures of the hive are some of the most valuable things that bees offer to humanity!

CHAPTER 9

Bee Pests and Diseases

An Ounce of Natural Prevention is Worth a Pound of Chemical Cure

The devastating effects of many bee diseases can be avoided by taking some simple steps to prevent them, and the key to disease prevention is clean, healthy, regularly inspected hives.

First, start with healthy, thriving bees. This sounds like a no-brainer, but how do you know if you are getting healthy bees to begin with?

Package Bees

If you are buying package bees from a distant apiary — ask pointed, thoughtful questions!

Some specific things to ask about include the following:

- what medications have been used to treat the source hives from which the packages were *shaken*
- whether the source hives have been engaged in migratory pollination
- what type of foundation is used in the source hives — *standard cell, small-cell*, plastic, etc.

If the answers to those questions aren't readily provided, you may want to keep shopping. Here are some of the reasons:

Bees that have been on the pollination circuit will have been artificially stimulated to build up the size of the colony early in the year, and they are medicated prophylactically with antibiotics due to their exposure to so many other bees while deployed on commercial pollination contracts. The hives are also worn and stressed from the travel involved. This background is not an ideal beginning for package bees. Please see Chapter 5, page 67 for a photo of a healthy bee package.

A better option is to purchase bees from source hives that stay home. The best source hives are hives that are not used for migratory pollination, and that were raised chemical-free on small-cell foundation (if on foundation at all) specifically for the purpose of making healthy packages.

Do your best to purchase from chemical-free apiaries. They do exist, and these beekeepers are dedicated folks. There are conflicts inherent in the methodology of package bees — it's not the most natural process — but a chemical-free apiary that is raising bees for packages on small-cell foundation is such a big step in the right direction. Supporting these apiaries is important!

Swarms

The more you know about the source hive of the swarm — the better. There is something positive to be said about the health of the colony if they were robust enough to swarm, but a swarm of bees raised on standard cell foundation and treated with chemicals may not be a strong as chemical-free, small-cell-raised bees. Please see Chapter 5, page 65 for a photo of a swarm.

You would essentially like to know the same things about a swarm as you would about a package — but so often when you collect a swarm, you are not able to talk to the beekeeper whose hive has swarmed, so it is unlikely that you will be able to get answers.

Cleanliness

Clean your hive tool and any other inspection-related equipment between hives, and especially if you find that you need to use your hive tool in another beekeeper's apiary. This one simple habit can reduce the vectoring of disease considerably.

Tidiness

Keep your apiary neat. Don't leave chunks of comb or propolis lying about on the ground to attract robbing bees or other critters.

Inspections

Inspect your hives regularly. I know that inspecting your bees, and learning what to look for can be intimidating—but the importance of recognizing signs of disease and other hive issues early is hard to overemphasize.

Learn Signs and Symptoms

Learn the signs and symptoms of bee diseases; know enough to know when you need help to identify a problem.

This book is not intended to provide deep insight into the pathology of all known bee diseases; there are many scientific books and research websites which excel at doing specifically that. *The Thinking Beekeeper* is meant to provide the small-scale, backyard beekeeper with enough good information that you can, first, recognize when something does not look right in your hive and, second, consider some natural options in the event you need to take action.

Terrible Things that Can Happen to Your Wonderful Bees

American Foulbrood

American Foulbrood, or AFB, is one of the most devastating and highly contagious of the diseases that affect honeybees. It is caused by the microscopic, spore-forming bacterium *Paenibacillus larvae*.

What Does It Look Like?

A disease of the brood, AFB is most readily identifiable by the sight of "sunken, perforated brood cappings." Instead of being slightly raised and a bit convex, the brood cell capping begins to turn dark, and then sink, becoming concave. The worker bees may puncture this capping.

American foulbrood is a seriously devastating, contagious disease of bee brood.

Credit: Virginia Williams, Image Number D827-1, US ARS Image Gallery. [online]. [cited September 7, 2012]. ars.usda.gov/is/graphics/photos/jul 07/d827-1.htm.

Honey and AFB

AFB really is the worst of the bee diseases, so this bears repeating: If you feed your untreated bees honey that was made by bees that were treated with antibiotics, it is possible that that honey could contain the spores that carry American Foulbrood. Your bees can contract American Foulbrood from that honey because your bees have not been treated with those same antibiotics. Feeding "white death" sugar is preferable to AFB!

The color of an infected larva inside the sealed cells changes from a bright healthy white to brown, darkening over time. The toothpick test is one method of confirming AFB. Puncture a capping with a toothpick — inserting the toothpick into the melted remains of the larva. If the toothpick is withdrawn with a brown ropy thread, you can be fairly certain that what you are seeing is AFB.

As the larva dries up, it turns into scale, which lies on the bottom of the brood cell and contains the spores that cause the disease.

These spores contaminate the honey, the comb and the hive equipment, and they live virtually forever.

What Can You Do About It?

Regulations concerning the treatment of AFB vary from place to place. Burning the hive is often considered the only truly effective solution. Antibiotics have been used but with undesirable after-effects, since their use is likely to lead to worse and recurring problems later.

The continued use of antibiotics also leads to the bacteria developing resistance and to contamination of the honey. Thinking beekeepers will inspect their hives regularly, being alert for the signs of AFB. If you suspect your hive has American Foulbrood, please contact your local apiarist or a relevant representative for assistance in confirming a diagnosis of American Foulbrood, and follow their guidance regarding disposal of both the bees and the hive equipment.

European Foulbrood

European foulbrood (EFB) is caused by the bacterium *Melissococcus plutonius*.

European foulbrood is also a bacterium affecting the brood, but unlike American Foulbrood, a larva infected with EFB usually turns brown and dies before the cell is capped. The bacterium does not create spores, but the bacteria left behind on the combs can be active for years, causing annual recurrences of the disease.

Infected larvae can die within four days after the eggs hatch. You are most likely to see EFB in the beginning of a season, when brood raising is most prevalent.

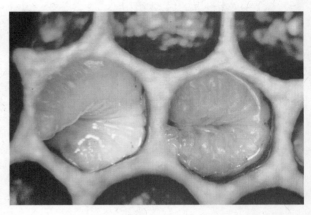

European foulbrood affects the larvae before the cells are capped.

Credit: Michael Wilson. Photo originally published on eXtension.org and used with permission. "European Foulbrood: A Bacterial Disease Affecting Honey Bee Brood." April 27, 2010. [online]. [cited September 7, 2012]. eXtension.org/pages/23693/european-foulbrood:-a-bacterial-disease-affecting-honey-bee-brood.

What Does It Look Like?

What you will see are dead larvae in the cells. The larvae is curled upward at one end and is brown or yellow and somewhat dry and rubbery.

What Can You Do About It?

European foulbrood is most likely to occur when the hive is stressed or unhealthy—sometimes due to weather or sporadic nectar availability, or in situations where the nurse bee population is too low to provide an adequate food supply to the brood. An otherwise healthy colony can usually survive European foulbrood.

Honeybee Viruses

In the US, there have been at least ten viruses reported that infect honeybees, including Kashmir bee virus (KBV), acute bee paralysis virus (ABPV), sacbrood virus (SBV), black queen cell virus (BQCV) and deformed wing virus (DWV).

However, only SBV and DWV cause any visible symptoms, so it's not known how important to bee health any of the other viruses actually are.

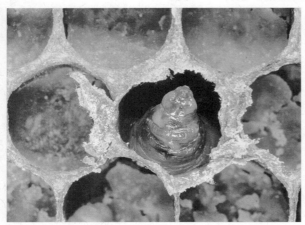

Sacbrood.

Credit: M.T. Frazier.

What Does It Look Like?

Sacbrood Virus (SBV) is a condition that affects brood and causes death of the larvae. Larvae with this virus fail to pupate, and fluid infected with the virus accumulates beneath their unshed skin. This forms a sac under their skin. The infected larvae change to a pale yellow color and shortly after death they dry out, forming a dark brown gondola-shaped scale.[1]

Deformed Wing Virus (DWV) causes abdominal and wing deformities on adult honeybees—and is most often seen in colonies heavily infested with varroa mites. Bees with DWV may have useless, shriveled wings, stunted abdomens and be discolored or paralyzed.

Bees showing DWV symptoms usually live less than 48 hours.[2]

Deformed wing virus.

Credit: M.T. Frazier.

What Can You Do About It?

No practical medical treatment for these viruses currently exists. Maintain good hygienic practices to avoid spreading these viruses. Awareness is imperative, and maintaining thriving healthy hives prevents bee viruses spreading further.

Chalkbrood

Ascosphaera apis is a fungal disease of bee larvae, which also propogates via spores. Chalkbrood is most commonly visible during wet springs.

What Does It Look Like?

The fungus will compete with the larvae for food, starving them to death. The fungus will then go on to consume the rest of the larvae's bodies, causing them to appear chalky white. The bees will often remove these mummified larvae, sometimes the remains can be found on the bottom of your hive.

What Can You Do About It?

Hives with chalkbrood usually recover on their own when dryer weather returns; beekeepers can help by increasing ventilation through the hive. Heavily infested combs should be removed from the hive and destroyed.

Wax Moth

Two types of wax moths concern beekeepers: *Achroia grisella* (lesser wax moth) and *Galeria mellonella* (greater wax moth). The beekeeper is more likely to see the adult moth than the larva—

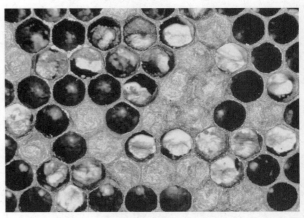

Top: The entrance to this beehive is littered with chalkbrood mummies that have been expelled from the hive by hygienic worker bees. Bottom: Chalkbrood mummies in brood cells.

Credit: (Top) Jeff Pettis, Image Number K8144-1, US ARS Image Gallery. [online]. [cited September 10, 2012]. ars.usda.gov/is/graphics/photos/300dpi/kesa/k8144-1.jpg; (Bottom) Photo by Professor M.V. Smith, University of Guelph. *Chalkbrood Disease*. WNCBees.org. [online]. [cited September 9, 2012]. wncbees.org/pests/Chalkbrood.cfm.

(Top) An adult greater wax moth; (middle) wax moth larva; (bottom) wax moth damage.

Credit: (Top) Image 5460334, Mark Dreiling, Retired, Bugwood.org. [online]. [cited September 10, 2012]. forestryimages.org/browse/detail .cfm?imgnum=5460334; (Middle and bottom) M.T. Frazier.

but it is the larva or caterpillar stage of both types that causes serious damage to the wax comb. The larvae can completely destroy wax comb while it is in storage, especially if the storage area is warm, dark and without good ventilation. Adult wax moths and larvae can transfer pathogens of other, more serious bee diseases such as foulbrood.

What Does It Look Like?

Wax moths are about ¾-inch long, grayish-brown, with a wingspan of 1¼ inches. The larvae look like large, maggot-like worms. At first glance, they are easily confused with the larvae of small hive beetles. (See box on opposite page.)

What Can You Do About It?

The bees themselves are the best form of control for these moths. They keep the populations at very low numbers, and it is only when they are not strong enough that the moth populations will drastically increase.

In the colder climates, storing beeswax combs in freezing winter temperatures will stop moth infestation. Wax moths need warm climates to thrive.

Small Hive Beetle

The small hive beetle (*Aethina tumida*) can be a destructive pest, causing damage to comb, stored honey and pollen in honeybee colonies. If a beetle infestation is serious enough, the bees may abscond.[3]

What Does It Look Like?

The larvae are cream-colored worms, a bit smaller than wax moth larvae. The beetles themselves are black and are about ¼-inch long. They are heavily armored, making them difficult to kill. The larvae are usually seen on comb; the pupae are found in the soil. Small hive beetle larvae may tunnel through combs of honey, feeding and defecating, causing damage and discoloration and fermenting the honey.

Hive beetles typically lay eggs in small cracks or crevices or directly on pollen and brood combs. The resulting larvae can destroy the combs.

What Can You Do About It?

Strong colonies and healthy bees are key, as small hive beetles are opportunists. Traps found in beekeeping catalogs can also be used — the ones typically filled with oil can be mounted on the interior side of a follower board.

> How can you tell the difference between small hive beetle larvae and wax moth larvae?
> - Beetle larvae frequently congregate in corners.
> - Beetle larvae never reach the size of mature wax moth larvae.
> - Beetle larvae have three pairs of jointed, "true" legs behind their heads.
> - Bodies of beetle larvae have tough exteriors.[4]

Small hive beetles are serious pests, especially in warm climates. The larvae are very destructive.

Credit: (Top) Jeffrey W. Lotz, Florida Department of Agriculture and Consumer Services, Bugwood.org, licensed under Creative Commons Attribution 3.0 License. [online]. [cited September 10, 2012]. bugwood. org/factsheets/small_hive_beetle.html; (Bottom) Florida Division of Plant Industry Archive, Florida Department of Agriculture and Consumer Services, Bugwood.org. [online]. [cited September 10, 2012]. bugwood.org/factsheets/small_hive_beetle.html.

Nosema

Two strains of Nosema disease trouble honeybees. *Nosema apis* and *Nosema ceranae* are each small microsporidian parasites that live in the digestive tract of honeybees.

Colonies in northern climates are more seriously affected than colonies in the south because of the increased amount of time bees are confined in the hive during the winter.

Nosema-stained hive.
Credit: Penn State Collection.

What Does It Look Like?

Nosema spores are microscopic in size. In addition, no symptoms are specific solely to nosema — the inability of bees to fly, defecation on combs or hive walls, and dead or dying bees on the ground in front of the hive may indicate nosema, or these may be symptoms of other bee diseases.

Infected bees often do not pay attention to typical brood-rearing tasks, and the hive population does not build up in spring. The bees are often unable to fly and may be seen crawling about with their wings spread out. Nosema reduces the life span of the bees and can seriously weaken or kill the colony.

What Can You Do About It?

I may sound like a broken record, but maintaining strong healthy hives is the best prevention for nosema. A thriving hive can often clear up a case of nosema when the weather breaks and flying weather starts. A weak, struggling hive is likely to succumb to the stress of nosema when added to its other early spring efforts to expand with brood building and foraging.

The use of antibiotics is prescribed by conventional beekeepers, but as we've discussed this comes with its risks and is generally outside the thinking beekeepers' paradigm.

Varroa Mites

Varroa destructor is an external parasitic mite that attacks honeybees. *Varroa destructor* can only replicate inside the cells of the pupae of a honeybee colony. The varroa mite attaches to the body of the bee, perforating its exoskeleton, leaving a permanent wound. The mite feeds by sucking the bee's hemolymph (blood). This makes the varroa mite a fantastic vector of disease and helps to contribute to the spread of viruses and other pathogens between bees and hives.

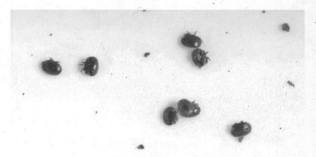

The varroa mite is a ubiquitous parasite of the honeybee.
Credit: (Top) Scott Bauer, Image Number K5111-10, US ARS Image Gallery. [online]. [cited September 10, 2012]. ars.usda .gov/is/graphics/photos/300dpi /kesa/k5111-10.jpg.; (Bottom) Image 5180012, Florida Division of Plant Industry Archive, Florida Department of Agriculture and Consumer Services, Bugwood .org. [online]. [cited September 10, 2012]. forestryimages.org /browse/detail.cfm?imgnum=518 0012.

What Does It Look Like?

Varroa mites are flat, oval-shaped and reddish brown with eight legs on the elongated side of the oval. They are about the size of the head of a pin. They are usually seen on the thorax and sometimes on the abdomen of a bee.

What Can You Do About It?

One of the best features of a top bar hive is that bees make all of their own natural wax, causing the size of the hexagonal cells of the honeycomb to be sized properly and apparently contributing to natural control of varroa mites. The likelihood of you ever seeing a mite load high enough to require action is small.

But it's imperative that beekeepers monitor for varroa mites. It's important to be certain that the mite load in your hive is low enough that no action needs to be taken. Because the mites can reproduce so quickly and are so talented at helping to spread disease, keeping varroa under control is probably the most important function of top bar beekeeping.

An adjustable, removeable bottom board, a bottom screened with ⅛-inch hardware cloth and a white *sticky board* (or *mite board*) are required to monitor for varroa mites. This is one of the design features built into the Gold Star Hive.

1. Remove the bottom board.
2. Using tacks or pushpins, attach the mite board to the bottom board.
3. Make the mite board sticky using vegetable shortening, or coconut oil, or a heavy coating of liquid vegetable oil.
4. Replace the bottom board in the lowered position, exposing the screened bottom.
5. Leave the mite board in place for three to five days to collect mites that have naturally fallen from your bees and through the screened bottom.

How to Dust with Powdered Sugar

Be sure to read the label on the powdered sugar before you purchase it at your local grocery store. It may contain cornstarch. Cornstarch is not good for the bees' digestive system, so the best way to avoid powdered sugar that contains cornstarch is often to take plain white granulated sugar and run it through a grinder to powder it yourself.

Then, fill a small flour sifter with the powdered sugar. Open your hive, move a bar or two out of your way as if you planned to do an inspection and sift the powdered sugar down between the bars onto your bees. Coat them fairly heavily. Move to the next bar, and the next, repeating the process until you have coated the entire hive of bees with a good layer of white powdered sugar.

At the end of that time, remove the board and count the total number of mites found on the mite board. Divide the total number of mites by the total number of days that you monitored. The result is known as *natural mite drop per day*. This number should be less than 30.

If the number of natural mite drop per day is higher than 30, then your next step should be to treat the hive with a dusting of powdered sugar.

Dusting with powdered sugar will cause the bees to groom themselves and each other in order to remove the powdered sugar — and will cause them to groom off most of the mites.

Treacheal Mites

Tracheal mites are parasites that live in the trachea (breathing tubes) of the bees.

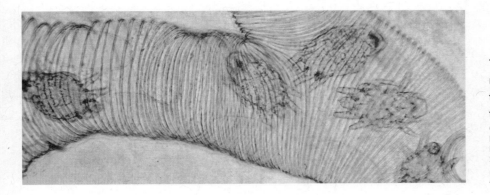

Tracheal mites are microscopic, living in the breathing tubes of the honeybee.

Credit: Lilia De Guzman, Image Number K5069-23, US ARS Image Gallery. [online]. [cited September 10, 2012]. ars.usda .gov/is/graphics/photos/dec04 /k5069-23.htm.

What Does It Look Like?

The mites are so tiny that they can only be seen with a microscope. Individual bees are believed to die because of the disruption caused to their breathing by the mites in their trachea, because of microorganisms entering their hemolymph (blood) through the damaged tracheae and from the loss of hemolymph.

John Skinner from the University of Tennessee explains, "The wings of infested bees are often unhooked…These bees are unable to fly and crawl about the hive entrance." Observers have seen infected bees leave colonies and die. Mite levels are usually highest in spring when there are fewer bees; mite levels are lowest in the summer when there are many bees. Tracheal mites increase when a hive is weak.[5]

If you see a lot of bees outside the hive in weather too cold to fly, as if they can't seem to get enough air inside the hive — that is one indication of tracheal mites.

What Can You Do About It?

Grease patties with sugar added have been used to treat tracheal mites. The bees consume the sugar and become coated with grease. The grease apparently impairs the mite's ability to reproduce or latch onto the bee.

Recipe for Grease Patties

Mix 1 ½ pounds of solid vegetable shortening with 4 pounds of granulated sugar until smooth. Form into hamburger-sized patties. The patties can be stored in the freezer, sealed in a plastic bag, until ready to use. Each patty should last at least a month.

Common Backyard Nuisance Pests

Mice

It was a number of years before I finally had a mouse make its way into a Gold Star top bar hive, but eventually it did happen. Mice are skilled acrobats and can squeeze their way into tiny, tiny places — places you absolutely wouldn't believe they could go. To prevent them from getting into the hive during the winter when they are in search of a warm place and food, install a piece of ¼-inch hardware cloth over any entrance that you are leaving open for the bees for the winter. Be sure to cork any other entrances. See the "Mice" section and the photo on page 126 in Chapter 7 for more discussion about safeguarding your hive against mice.

An end entrance hive can be protected from mice with a wooden *entrance reducer* — a piece of wood cut to fit tightly in the entrance, allowing only a small hole for the bees to enter and exit.

I also recommend filling the space above the top bars inside a gable roof with something such as bagged insulation, as this helps to discourage mice from moving into the attic space above the bars.

Bears

Happily, bears are not common in the urban backyard, but they are very common in rural areas, so they deserve mention here. Contrary to popular children's stories about bears loving honey, bears are actually more interested in the protein of the honeybee larvae in a hive. While the honey is a sweet addition and the smell probably helps to attract the bear, the original lure was likely not the honey.

There are few deterrents to bears aside from a strongly charged electric fence. Placing a piece of bacon or a glob of peanut butter on the hot wire of an electric fence to attract and then "shock train" the bear will usually deter them from attempting to reach the hive.

Skunks, Raccoons, Opossums

Skunks and other small mammals that forage at night, such as raccoons and opossums, like to eat bees, but they generally don't like to get stung in the face or on their underbelly. Skunks in particular will scratch at the wood of the hive, causing the guard bees to come out to challenge the source of the noise. When the bees come out, the skunk grabs and rolls the bees and then eats them.

If your hive is elevated on legs or on a stand, this causes the skunk to have to stand up to get to the entrance of the hive, exposing its tender underbelly. This lessens the likelihood of a skunk bothering the hive.

If your top bar hive suddenly gets very grouchy, look for signs of these animals scratching at the wood of the hive — and elevate it.

Chickens

Yes, it's true that chickens will eat bees. This is another instance where having the hive elevated on legs will usually avoid problems.

Ants

Unless the ants in question are carpenter ants — and are literally consuming the wood of the hive — ants are probably more of an annoyance to the beekeeper than to the bees. Usually they are only looking for a dark crevice to lay their eggs, and so you will often see them between the observation window's shutter and the glass.

Brushing them away with your bee brush is all the maintenance really required — though you may find that a sprinkling of borax around the base of the hive legs or into the space between glass and shutter will encourage them to lay their eggs elsewhere.

If they are biting ants and you are at the hive in order to do an inspection, I recommend that you leave them where they are, perform your hive inspection and then as your last act, brush them away. Otherwise you may find yourself standing in stinging ants while you inspect your hive — an experience that I can assure you, you will not enjoy!

This has been an overview of the most common bee diseases and pests. There are many good sources for more detailed information and deeper research. See Appendix B for resource lists.

Afterword

And the Lord inspired the bee, saying: "Take your habitations in the mountains and in the trees and in what they erect. Then, eat of all fruits and follow the ways of your Lord made easy (for you)." There comes forth from their bellies a drink of varying colors wherein is healing for men. Verily in this is indeed a sign for people who think.

« Quran 16:68–69[1] »

Early Enthusiasm

Early on in my beekeeping career, I can remember being fervently convinced that treatment-free beekeeping in top bar hives was the best, was in fact the only good beekeeping in existence. I insisted that I was never, ever, ever going to treat my hives with chemicals. I was never, ever going to feed them white processed sugar; I was never again going to keep bees in square boxes. I was sure that my bees would prevail against all odds by gathering the lovely, healthful nectar from the pretty flowers near my home, and that would be enough to keep them all fed and healthy and they would all live forever, or at least happily ever after.

In short, I was a zealot!

I was prepared to wage a one-woman holy war against those evil migratory pollinators who treated their hives with chemicals and then dragged their poor bees, boxes stacked onto pallets, pallets forklifted onto tractor-trailers, all over the United States.

I was prepared to rant and rave against toxic chemical treatments and antibiotics in beehives, against feeding honeybees sugar and seriously against harvesting their honey, only to replace it with gallons of high fructose corn syrup.

I was astonished by how callous and how casual conventional beekeepers were about manipulating their bees.

It made me crazy! And truthfully, it still does.

Thwarting the raising of drones, controlling the gender of bees, altering the size of the bee and the length of its gestation period through the use of pre-printed foundation—that still seems like a deep level of manipulation to me. That's why I'm such an advocate for natural wax—and against the use of foundation.

The use of chemicals in a beehive—toxic insecticides placed directly into the nest of this precious and important insect—still strikes me as counterintuitive. First there's the direct effect of the chemicals on the adult bees; then there's the issue of the beeswax comb absorbing, like a sponge, those persistent toxins, affecting the queen, the colony's future brood and their food stores. And frighteningly, the fact that these chemicals survive the rendering process means that they persist even into the production of new sheets of foundation.

Seen from a larger perspective, propping up sick bees with the use of toxic chemicals really means that we are perpetuating weak bees—and not letting nature follow its course of supporting the healthiest, strongest bees—the bees that are "tough enough to take it." That's why I still believe that the best course of action is to never put anything in a beehive but bees.

But time has passed from those early over-zealous novice days...and during that time I learned a bit more about beekeeping. I met some of those "evil migratory beekeepers"... I lost some bees of my own... Soon it became obvious that there were issues at stake that were bigger than they appeared at first blush.

I began to expand my thinking a bit.

Making the Connection...

This expansion occurred for me largely because Colony Collapse Disorder (CCD) was the plague du jour when I became a beekeeper, and the research being done into CCD and its possible causes began to draw a great deal of public awareness to some other very important things that were also happening. These things were all closely related, but they weren't all specifically about bees and beekeeping.

In the TEDxDirigo talk I gave on September 10, 2011 — "Making the Connection: Honeybees, Food, and You" — I did my best to highlight the connection between our broken food system, our current agricultural methods, government policies, systemic pesticides, GMOs and the effects that all these things were having on honeybees. These things weren't really about beekeeping — and yet, they were.

The analogy of the honeybee as a modern day "canary in a coal mine" was suddenly completely apropos, and drove home the point that CCD was not just another bee pest or disease — but a symptom of much larger, interrelated problems.

Speaking of the Government...

I've devoted a fair bit of this book to vague references to "the government," and I've complained about the ineffectiveness of some governmental agencies, especially as they connect to our food system and the environment. But in all fairness, not everything that issues forth in the name of the US government is a bad thing. Sometimes they are on the right track from the very beginning.

Let me share with you a personal story that is near and dear to my heart.

In the spring of 2009 the Obama administration, with a nudge from the On Day One contest and the viral campaign "Eat the View," instigated by Roger Doiron of Kitchen Gardeners International,[2] planted an organic garden on the White House lawn. The simple act of driving a shovel through the manicured green desert surrounding the residence of our nation's leader and turning over the soil was groundbreaking in more ways than one. Organic food began to matter in a completely different way.

Caption: Thanks to Charlie Brandts for a wonderful visit to the White House garden and the First Bees.
Credit: Jim Fowler.

Not long after, Charlie Brandts became the "First Beekeeper," tending a hive of bees near the organic garden and illustrating another ground-breaking connection at the White House. In October of 2009, I was privileged to visit the White House garden and its bees, and I talked with Charlie about treatment-free beekeeping, top bar hives and what it was like to be known as the "First Beekeeper."

Since those early days, First Lady Michelle Obama's Let's Move! campaign has been working to solve the childhood obesity problem,[3] an epidemic brought on by changes in lifestyle since 1970 — changes that have lowered the amount of exercise that children get, increased the number and portion sizes of snacks, and increased the total number of calories people in the US consume by 31%. If we don't solve this problem, a third of all children born since the year 2000 will be diagnosed with diabetes at some point in their lives. Many others will face other chronic obesity-related health problems like heart disease, high blood pressure, cancer and asthma.

Growing an organic garden is such a great first step to resolving the brokenness of our food system, salvaging our health and the health of our children and restoring the crucial balance of the natural systems that we all depend on. That bees were included as a part of the White House organic garden illustrates the connectedness in a very big way.

What a high point for me! The creation of this garden really strengthened my conviction that, fork by fork and garden by garden, we can make the changes our food system needs. My heartfelt thanks to the Obamas for these two simple, meaningful acts.

Shifting to a Bigger Paradigm

In the documentary film *Vanishing of the Bees*, Michael Pollan comments "We don't have to wait for the government." He goes on to say that we can all "vote three times a day — with our forks!" So while the government's behavior may be confusing—growing an organic garden on the White House lawn at the same time that the EPA approves the use of neonicotinoid pesticides, GMOs and other things that are damaging the fragile workings of our food system—we still have a great deal of say in the matter.

The point that Pollan makes is important: The choices you make when you spend your food dollar have a tremendous impact.

Consider: The behavior of any corporation is dictated by money. The corporation follows trends; it tries to anticipate how consumers will allocate their hard-earned dollars. Spending your money is very similar to how enthusiastically you applaud after a live performance.

When you spend your money on conventionally grown food—food produced by unsustainable methods, food that is unhealthy for you, detrimental to the planet and bad for bees—it is as if you are applauding those farming methods, maybe even giving them a standing ovation and certainly encouraging them to continue their performance.

So be aware of that, and if that is not the message that you mean to convey, then change where you spend your food dollars. We don't have to wait for the government to tell us to do that. We can make that change—and let the government catch up!

But Let's Go Back to Talking about Bees, Shall We?

By shaking us up with Colony Collapse Disorder, honeybees have done their part as our early warning system. Now it's up to us to pay attention, and act on that warning. We have to learn to see ourselves as the shifters of the existing paradigm. We must, as Gandhi said, "Be the change we want to see in the world."

Repairing our agriculture practices, practicing treatment-free beekeeping, eschewing the use of pesticides, spending our food dollars appropriately and restoring respect for all of the natural systems required for the planet to sustain itself and us: These are the ways forward.

And after taking those steps, we will soon realize that we didn't have to find a new cure for CCD...

We just had to quit causing it.

Endnotes

Chapter 1: How Did We Get Here From There?

1. L. L. Langstroth. *Langstroth on the Hive and the Honey-Bee: A Bee Keeper's Manual.* Hopkins, Bridgman & Company, 1853, p. 15. [online]. [cited August 27, 2012]. gutenberg.org/files/24583/24583-h/24583-h.htm.

2. Ibid.

as Braatz. *Bees.* Steiner Press, 1998.

Dilemma: A Natural History in Four Meals.

and Michael Pollan. *The Botany of Desire:*
Random House, 2001.

Vax

te and the honeybee are documented in a video
USDA Agricultural Research Service. Jeff Harris.
ee and Varroa Mite." [online]. [cited July 30, 2012].
content.asp?type=wax or video.google.com/video
50616&hl=en.

: "A Test for Sub-Lethal Effects of Some Com-
Year Two." 2010 American Bee Research Confer-
]. [cited August 29, 2012]. eXtension.org/pages
ib-acute-effects-of-some-commonly-used-bee

3. Christopher A. Mullin et al. "High Levels of Miticides and Agrochemicals in North American Apiaries: Implications for Honey Bee Health." PLoS ONE, March 19, 2010. [online]. [cited May 3, 2012]. plosone.org/article/info%3Adoi%2 F10.1371%2Fjournal.pone.0009754.

4. Reed M. Johnson. "When Varroacides Interact." *American Bee Journal* and *Bee Culture*, December 2009. [online]. [cited May 3, 2012]. beeccdcap.uga.edu/docu ments/CAPArticle2.html.

5. Cornell Extension Toxology Network (EXTOXNET). [online]. [cited May 3, 2012]: Coumaphos Pesticide Information Profile: pmep.cce.cornell.edu/profiles /extoxnet/carbaryl-dicrotophos/coumaphos-ext.html; Fluvalinate Pesticide Information Profile: pmep.cce.cornell.edu/profiles/extoxnet/dienochlor-glyphos ate/fluvalinate-ext.html.

Chapter 4: The Top Bar Hive

1. Thomas D. Seeley. *Honeybee Democracy*. Princeton, 2010.
2. For more on the importance of cell size, please see Chapter 2.
3. Dennis Murrell. "Condensation." Bee Natural Guy blog. [online]. [cited August 1, 2012]. beenaturalguy.com/observations/condensation/.
4. One such source is David Heaf. *The Bee-friendly Beekeeper: A Sustainable Approach*. Northern Bee Books, 2011.

Chapter 5: On Getting Started with Your Own Top Bar Hive

1. Your can find a mentor by consulting Bee Culture Magazine. *Who's Who in North American Beekeeping*. [online]. [cited August 30, 2012]. beeculture.com/content /whoswho/.
2. For more information on removing bees from buildings, here's a helpful book: Cindy Bee and Bill Owens. *Honey Bee Removal: A Step by Step Guide*. A.I. Root Co., 2010.
3. Gunther Hauk. *Toward Saving the Honeybee*. Steiner, 2003, p. 58.
4. Pesticide Action Network North America. *Pesticides on Food*. [online]. [cited August 3, 2012]. panna.org/issues/food-agriculture/pesticides-on-food.

Chapter 6: Inspections

1. Visit the beetight.com website. There's an app for that!
2. See National Library of Scotland, The Moir Rare Book Collection. "'Tanging' the bees." [online]. [cited August 8, 2012]. digital.nls.uk/moir/tanging.html.

Chapter 7: Overwintering Your Top Bar Hive

1. Bee Informed Partnership. *About*. [online]. [cited August 4, 2012]. beeinformed .org/about/
2. Ibid.
3. Wolf Creek Apiaries, LLC. [online]. [cited August 6, 2012]. wolfcreekbees.com.
4. info@goldstarhoneybees.com.

Chapter 8: Treasures of the Hive

1. PRLog. "The Hydrogen Peroxide Producing Capacity of Honey." April 29, 2009. [online]. [cited June 29, 2012].prlog.org/10227103-the-hydrogen-peroxide-producing-capacity-of-honey.html.
2. The American Apitherapy Society Inc. [online]. [cited June 29, 2012]. apitherapy .org/.

Chapter 9: Bee Pests and Diseases

1. Elvira Grabensteiner et al. "Sacbrood Virus of the Honeybee (Apis mellifera): Rapid Identification and Phylogenetic Analysis Using Reverse Transcription-PCR." *Clinical and Diagnostic Laboratory Immunology*, Vol 8#1 (January 2001), pp. 93–104. [online]. [cited August 13, 2012]. ncbi.nlm.nih.gov/pmc/articles /PMC96016/.

2. Wikipedia. *Deformed wing virus*. [online]. [cited August 13, 2012]. en.wikipedia .org/wiki/K-wing_deformity.

3. Wikipedia. *Small hive beetle*. [online]. [cited August 14, 2012]. en.wikipedia.org /wiki/Small_hive_beetle.

4. John Skinner, University of Tennessee. *How can I tell the difference between small hive beetle larvae and wax moth larvae?* eXtension.org, May 14, 2012. [online]. [cited September 9, 2012]. eXtension.org/pages/44111/how-can-i-tell-the-differ ence-between-small-hive-beetle-larvae-and-wax-moth-larvae.

5. John Skinner, University of Tennessee. *What are the symptoms of tracheal mite infestation in honey bee colonies?* eXtension, org, November 10, 2009. [online]. [cited September 9, 2012]. eXtension.org/pages/44108/what-are-the-symptoms -of-tracheal-mite-infestation-in-honey-bee-colonies.

Afterword

1. Islamweb. *The miracle of honey as an alternative medicine*. April 24, 2011. [online]. [cited September 2, 2012]. islamweb.net/emainpage/index.php?page=articles &id=138095.

2. Kitchen Gardeners.org. *This Lawn is Your Lawn*. [online]. [cited September 2, 2012]. youtu.be/dkMVTM0Gszw.

3. Let's Move! *Gardening Guide*. [online]. [cited September 4, 2012]. letsmove.gov /gardening-guide.

Glossary

abscond (*v*)/**absconding:** when a colony of bees abandons a hive because of some condition in the hive itself — such as an infestation of pests, e.g. small hive beetle — or because of a condition in the environment, e.g. a constant irritating light or noise

alarm pheromone: there are two types of alarm pheromone. One is released when a bee stings — which alerts other bees to the location of a threat and can excite them to sting as well. The smell of this one is similar to the smell of bananas, so it's good to avoid eating bananas before working your hive! The other pheromone is repellent and used to warn of potential enemies and robbing bees. Smoke can be used to mask these alarm pheromones

anaphylaxis/anaphylactic shock: a fast-occurring, life-threatening, systemic allergic reaction

anecdotal evidence: evidence from anecdotes or stories told by others. Not always reliable, but it abounds in beekeeping!

acaracide: a type of pesticide used to kill mites

attendant bees: bees that are included inside a queen cage when purchasing a package of bees, or an individual queen bee. They attend to the queen — feeding and grooming her

backyard beekeeping: small-scale beekeeping done in one's own backyard, with emphasis on natural methods and working with the bees' natural systems. The antithesis of industrial beekeeping, it typically eschews the use of chemical treatments and antibiotics in the hive

ball (*v*)/**balling:** bees cluster around the original queen in a very tight ball, increasing the temperature within the ball to the point where she is killed by the heat they generate

barbed stinger: a honeybee's stinger is barbed, much like a fishhook. This means that once it penetrates skin or clothing, it lodges there.

bee bowl: the open space inside a top bar hive where the bees will be poured or installed. The rest of the hive should have its top bars all in place so that bees fall only into the bowl part.

bee ladder: any device that provides a surface for bees to walk on to approach a liquid food source, designed to prevent them from drowning; often made of a piece of hardware cloth bent into a checkmark, or sliding board shape

bee space: ⅜ of an inch, the amount of space required between combs — large enough for the bees to walk between the combs.

beeswax: the wax secreted by honeybees from glands in their abdomens, used to build their comb

bonk (*v*)/bonking bees: an amusing and not highly scientific phrase describing the firm tap given to a container of bees when hiving them. This movement breaks the cluster of bees into a heap, which can then be poured into the hive.

brood: the egg, larval and pupal stages of the honeybee — in other words, baby bees!

brood comb: the sheets of beeswax nesting material in a beehive where the baby bees are raised. See comb.

burr comb: random wax built between combs or between hive parts

candy: a sugar concoction thick enough that holes can be plugged with it

cap (*v*): the process of the bees building a "lid" or covering over each cell of ripe honey or brood

capped honey cells: honey cells sealed with the wax lid made by the bees

capped brood cells: the pupal stage of the bees' gestation period, wherein the cells have a fibrous cap or covering sealing the pupae in the cell

caste: term used to designate the types of bee found in the hive: queen, worker, drone

catenary curve: a mathematical term for the rounded shape of natural honeycomb, also the curve defined by hanging chain theory

cavity: an enclosed space or chamber, specifically one of sufficient volume to house a colony of honeybees

cavity nesting: the habit, typical of honeybees and certain other insects and animals, of occupying a cavity and building their nest inside it

CCD: Colony Collapse Disorder

center side entrance: a descriptive term for a top bar hive with its entrance(s) in the center of the long side of the hive

cleansing flight: a flight taken by honeybees after a period of being dormant in the hive, especially in winter, during which they defecate after being closed in for an extended period of time

clipping: an attempt to thwart the swarming mechanism. Tiny scissors are used to cut a queen bee's wings so that she is unable to fly and leave the hive even when the colony is prepared to reproduce. This method is not generally successful in keeping the hive from swarming, as one of the virgin queens produced as part of this process leads the swarm instead.

Colony Collapse Disorder: the name given in 2006 to the phenomenon that garnered the attention of beekeepers and researchers everywhere: Entire colonies of honeybees simply disappear, leaving behind their brood and food stores, but no dead bees. This is in striking contrast to the bees' natural behavior of defending brood and food by stinging, being loath to abandon either.

colony of lost boys: a colloquial name for a queenless hive with a laying worker, as the number of drone or boy bees soon begins to exceed the number of worker or girl bees, sending the whole hive into a downward spiral

comb: the sheets or panels, consisting of hexagonal cells of beeswax, that make up the nesting material inside a honeybee hive. Also referred to as brood comb or honeycomb depending on its purpose in the hive.

cross-comb: the term used to describe the bees building comb off the comb guides provided in a top bar hive and instead building across multiple top bars, locking them together and preventing the removal and inspection of individual bars. This is to be avoided in the interests of disease prevention, and because movable comb hives are legally required in many locales.

crush and strain: the process in which combs containing honey are removed from the hive, placed in a strainer and crushed to break open the cells allowing the honey to drain out of the crushed beeswax and through the strainer before being bottled

cut comb honey: sections of honeycomb containing honey still sealed inside the hexagonal cells

day count: the number of days elapsed since a colony of bees was installed into a hive

DCA: drone congregation area

destructive harvest: a method of harvesting honey where the entire colony, including the bees, is destroyed

draw (v): the process of building beeswax out to the full depth of the hexagonal cells of honeycomb, whether done by the bees with beeswax or by a manufacturer of fully drawn plastic foundation

drive-through window: a hole drilled through a follower board that contains the bee colony inside the hive, allowing access to the feeder area of the hive

drone bee: the male honeybee

drone congregation area: a place where drones congregate waiting for queen bees that fly through this area in order to mate

egg: the first three to four days of the gestation period of a honeybee

emergency queen replacement: a situation where the bees work to replace the queen due to her sudden loss. It requires the presence of larvae three to four days old.

end entrance hive: a descriptive term for a top bar hive with its entrance(s) on one end of the hive, whether on the end panel (typically a slot running the width of the end panel) or at one end of the long side

entrance: an opening in the hive where the bees enter and exit

entrance reducer: in an end entrance hive, a piece of wood that is cut to fit securely in the entrance, with only a small hole for the bees to enter and exit. In a center side entrance hive with several round holes for entrances, corks can be used to close some of the holes, leaving only one open in winter, or to prevent robbing.

extractor: a mechanical device used to remove honey from honeycomb

false end: See follower board.

feed can: a can filled with syrup and having small holes in the bottom, installed inside a bee package to provide food for the bees while in transit

feeder: a dispenser used to feed sugar syrup to bees

festoon: the way bees hang in a gentle curve inside the hive to create natural honeycomb

fixed comb hive: a beehive where the individual combs are fixed, or attached, to the interior of the cavity or container in which the bees are living. See also log gum.

follower board: a board or boards designed to contain the bees within an area smaller than the entire hive. Used when starting a hive to relieve a small starter colony from having to heat the entire hive.

forage (*n*): bee food

forage (*v*): the process of bees searching for food

foundation: sheets of wax or plastic embossed with hexagons that are installed in the frames of a conventional or Langstroth beehive and intended as a guide to the bees in building their comb

foundationless hive: any type of beehive that does not use foundation, but allows the bees to make their own natural beeswax comb

frame: the rectangular structure found inside conventional hives that surrounds the beeswax combs; it typically contains a sheet of wax foundation. See top bar.

fully drawn foundation: foundation which has the hexagonal cells completely constructed or drawn out from the mid-rib. May be made of wax or plastic.

GPS: global positioning system

growing season: the time of spring/summer/fall when the flowering plants that provide bee food are growing

hanging chain: See catenary curve.

hatch (*v*): to evolve from egg to larva

hitchhiker: slang term for a honeybee that hitches a ride on the outside of a package of bees

hive: the home of a colony of honeybees, also the colony itself

hive mind: the ability of a colony of honeybees to make decisions as one cohesive unit

hobby beekeeping: See backyard beekeeping.

Hoffman frames: the name for the rectangular frames found inside Langstroth hives

honey: the sweet amber-colored substance produced by honeybees from the nectar and pollen of flowers

honeybee: In this book, honeybee means the species *apis mellifera*, the western or European honeybee. A note about the spelling of the word: Many dictionaries list "honeybee" as one word. An entomologist would more likely use the two-word naming convention "honey bee" as established by the Entomological Society of America. So, technically, both are correct. In this book, we are using the dictionary spelling and not the ESA naming convention (Entomological Society of America. *Common Names of Insects and Related Organisms.* [online]. [cited June 6, 2012]. entsoc.org/pubs/common_names)

honeycomb: the sheets of beeswax nesting material in a beehive where honey is stored

honey gate: a valve installed in a honey collection bucket that can be opened and closed to allow easy bottling

honey hunting: the practice of gathering honey from wild hives in nature instead of from managed artificial hives

honey super: a box designed for use with a Langstroth hive for the harvest of honey; generally of shallow depth for ease in lifting

iconoclast: a person who attacks or abandons long-held, traditional beliefs in favor of new ideas. I use this term to describe *thinking beekeepers*, you who are reading this book!

increase (*v*)/increasing: See split.

industrial beekeeping: beekeeping practices that were developed to dovetail with industrial agriculture systems that utilize monoculture growing practices. It includes the practice of migratory pollination and typically supports the use of chemical treatments and antibiotics in the hive.

landing board: a flat surface in front of the entrance to a beehive provided as a place for bees to land when returning to the hive, considered by some to be a crucial component of hive design, and by others to be unnecessary

Langstroth hive: the square-box hive with removable frames containing the bees' honeycomb that was patented by Reverend Lorenzo Lorraine Langstroth in 1853. Typically utilizes sheets of pre-printed wax foundation in the frames as comb guides. The most common type of beehive in use in the US in 2012

larva (*s*)/larvae (*pl*): the developmental stage between egg and pupa. A larva looks like a glistening white, c-shaped grub or worm

laying worker: a worker bee that has begun to lay eggs in a queenless hive, when the bees' ovaries are no longer suppressed by the queen's pheromone

locked up: the results of cross-combing inside a foundationless beehive, where the wax is not drawn in alignment with the top bars but across several of them, locking the bars together and creating a fixed comb hive that cannot effectively be inspected

log gum: the name given to a hive that consisted of a section of tree containing a colony of honeybees, which was removed by the beekeeper and placed in an apiary. Often these colonies were found in gum trees, which naturally rot from the inside out, creating an ideal cavity for honeybees to nest in. This type of hive is considered a fixed comb hive.

managed beekeeping: the practice of keeping bees in artificial hives that can be managed or manipulated by humans

marked queen: a queen bee who has had a tiny dot of paint applied to her thorax

mid-season shift: a management technique specific to center side entrance top bar hives. The entire colony is shifted into the empty one third of the hive in order for the bees to continue building their comb in the direction they began.

migratory pollination: the practice of moving beehives in order to pollinate crops necessitated by the practice of monoculture agriculture

mite board: a sheet of reusable white plastic, made sticky with oil or shortening, and installed beneath the screen bottom of a beehive to monitor for the presence of varroa mites; also called sticky board

monoculture: the industrial agricultural practice of cultivating huge areas of farmland with but a single crop. This is devastating to the natural balance of pests in the area and causes an increase in the use of pesticides since a pest that thrives on the crop being grown is easily able to wipe out the entire area

mouse guard: a device installed in or over the entrance of a beehive in the fall to prevent mice from getting into the hive to nest

movable comb: comb that can be removed from the hive and inspected for the purpose of disease detection and prevention, legally required in the US today

nadir (*v*): to add a hive box to the bottom of a hive stack. This is typically done with Warre hives.

natural cell size: the size of the hexagonal cells in honeycomb that bees draw when they make their own natural wax comb, typically 4.9 mm.

natural mite drop per day: a mite count method used when monitoring the level of varroa mite infestation in a beehive. It is calculated by dividing the total number of mites found on a sticky board by the total number of days monitored

nectar: the sweet liquid produced by flowers to attract bees who help pollinate the plants that produce the nectar

nectar dearth: a condition where weather (temperature, season, drought, frost) causes plants to produce little or no nectar

nuc: shorthand for the term nucleus, a starter hive consisting of hive equipment, drawn comb with brood and stores as well as bees. As of this writing, nucs are commercially available for Langstroth hives; they are much less frequently available for top bar hives.

paradigm shift: an overarching change in worldview which affects theories and methodology. This is presently occurring in agriculture, where there is a shift away from the use of toxic pesticides and fertilizers and towards organic methods. This shift has been instigated by people who *think*, and it is happening now!

persistent: existing or continuing for a long time; degraded only very slowly by the environment; specifically describes pesticides which have survived the recycling process of being melted down to liquid and milled into fresh sheets of foundation

pheromone: an airborne chemical factor produced by honeybees that triggers a response from the bees in the hive

plumb: vertical

pollen: the grainy or powdery substance put out by the male part of a plant which, when transferred to the female parts, causes pollination or fertilization of the plant, which is then able to set fruit

propolis: a glue-like substance made by the bees from the resin of trees

pupa (*s*) / pupae (*pl*): the stage of the honeybee's gestation period between larva and adult. This stage occurs inside a thin silk cocoon enclosed in a capped brood cell.

queen bee: the reproductive female bee in a colony of honeybees

queen cage: a small container or device used to contain a queen bee

queen cells: the vertical cells in which queen bees are raised

queen cup: the small wax cup which is a precursor to a queen cell

queenlessness: the state in a beehive where the queen has been lost to death or injury. The bees are very attuned to the existence of the queen, and their behavior changes radically when they are queenless.

queen pheromone: the substance emitted by a queen bee that is responsible for the cohesive nature of a colony of honeybees

queenright: the state in a beehive where the queen is alive and well

remainder hive: the bees, comb, brood and food left behind in the original hive when a colony swarms

ripe honey: honey that has been evaporated down to the proper moisture level and capped with a wax lid by the bees

shake/shaken: the process of gathering bees to make package bees, by shaking them from the frames of existing hives into a collector box and from there into packages

skep: a dome-shaped beehive typically made from straw or wicker. The bees build their combs attached to the inside of the dome. This type of hive is considered a fixed comb hive.

slum gum: the pupal cocoons and other debris left behind after rendering beeswax

small-cell foundation: wax foundation sheets with embossed hexagons measuring 4.9 mm

smooth stinger: a type of stinger with no barb, typical of yellow jackets, wasps and hornets enabling these insects to sting their victims multiple times

spacer: the thin strips of wood provided in a Gold Star hive to allow the beekeeper to space the top bars further apart when the bees begin to draw honeycomb — which is typically drawn thicker than brood comb

split (*n*): a hive produced by dividing an existing colony into two hives, especially when the original colony is preparing to swarm

split (*v*)/splitting: dividing one colony of bees into two hives, also known as increasing

spring start-up: the beginning of beekeeping season when new hives are started, and overwintered hives are unwrapped in anticipation of the spring

standard cell foundation: wax foundation sheets with embossed hexagons measuring 5.4 mm

sticky board: a sheet of reusable white plastic, made sticky with oil or shortening, and installed beneath the screen bottom of a beehive to monitor for the presence of varroa mites; also called mite board

sublethal effects: effects of treatments or manipulations that damage but do not immediately kill the honeybee — a serious concern when studying the long-term effects of systemic pesticides and other chemical treatments

super (*n*): an additional Langstroth hive box

super (*v*): to add an additional box to the top of a hive stack, as is typically done with Langstroth hives

supersedure: the process by which a colony of honeybees replaces a failing queen

swarm: the method by which a colony of honeybees reproduces. It involves the raising of drone bees, new queens in swarm cells and then the departure of roughly half the colony accompanied by a queen bee in search of a new cavity in which to build a nest.

swarm cell: the type of cell in which a new queen bee is raised when a colony of bees is preparing to swarm; vertically oriented, and similar in appearance to a peanut shell

tang (*v*)/tanging: the practice of beating rhythmically on a metal pan or drum when bees are swarming, thought to cause the swarm to settle lower in a tree

thinking beekeeper: the thoughtful kind of person — like you — who believes that no chemical treatments and no antibiotics should be used in the keeping of bees, that only bees should be put into a beehive

top bar: the bars placed across the cavity of a top bar hive, which will act as comb guides to encourage the bees to build natural beeswax comb from each bar, creating a movable comb hive. See frame.

top bar hive: a horizontal beehive with bars, known as top bars, that rest across the top of the hive cavity, each bar touching the next, from which the bees build their natural beeswax comb

varroa mite: a debilitating external parasite of the honeybee. Varroa mites feed off the body fluids of honeybees and vector harmful diseases.

wall-to-wall bees: a full, thriving hive of bees, with drawn comb on each bar in the hive

wax: beeswax

windshield wiper technique: a procedure for separating top bars in which you carefully run the blade of a top bar hive tool down the inside wall of the hive until the handle comes to rest upon the top edge of the hive body. Then, using the handle as a fulcrum, swivel the blade upward, toward the bottom of the top bars, thus safely removing any wax that may be attaching the comb to the side of the hive. Important to do this by cutting upward to prevent tearing comb from the bar.

worker bee: a female honeybee

work (v)/working the hive: the process of opening and inspecting a hive

Appendix A:
Sample Hive Inspection Diagram

Beekeepers: Copy this diagram for use in your Bee Log. Use the legend to note the contents of each bar. Be sure to date each entry, identify the hive if you have more than one, and add any notes about weather, hive temperment, swarm preparations, the age of the hive, etc.

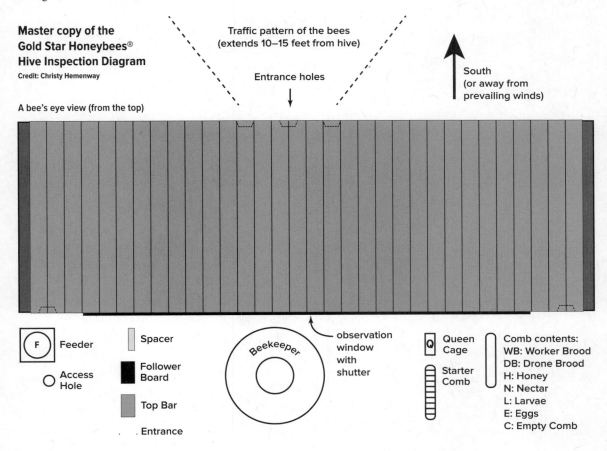

**Master copy of the
Gold Star Honeybees®
Hive Inspection Diagram**

Credit: Christy Hemenway

A bee's eye view (from the top)

Traffic pattern of the bees
(extends 10–15 feet from hive)

Entrance holes

South
(or away from
prevailing winds)

F Feeder

Access Hole

Spacer

Follower Board

Top Bar

Entrance

Beekeeper

observation window with shutter

Q Queen Cage

Starter Comb

Comb contents:
WB: Worker Brood
DB: Drone Brood
H: Honey
N: Nectar
L: Larvae
E: Eggs
C: Empty Comb

Appendix B:
Bee Resources

Books

Blackiston, Howland. *Beekeeping for Dummies*, 2nd ed. Wiley, 2009.

Bush, Michael. *The Practical Beekeeper: Beekeeping Naturally, Volumes I, II and III.* X-STAR Publishing, 2011.

Chandler, Phil. *The Barefoot Beekeeper*, 3rd ed. Lulu, 2009.

Conrad, Ross. *Natural Beekeeping: Organic Approaches to Modern Apiculture.* Chelsea Green, 2007.

Crane, Eva. *The World History of Beekeeping and Honey Hunting.* Routledge, 1999.

Crowder, Les, and Heather Harrell. *Top-Bar Beekeeping: Organic Practices for Honeybee Health.* Chelsea Green, 2012.

Delaplane, Keith. *First Lessons in Beekeeping.* Dadant & Sons, 2007.

Flottum, Kim. *The Backyard Beekeeper: An Absolute Beginner's Guide to Keeping Bees in your Yard and Garden*, rev. and expanded. Quarry, 2010.

Heaf, David. *The Bee-friendly Beekeeper: A Sustainable Approach.* Northern Bee Books, 2011.

Huber, Francis, translated C. P. Dadant, transcribed Michael Bush. *Huber's New Observations Upon Bees The Complete Volumes I and II.* Originally published 1789–1814. New edition with enhancements, X-Star Publishing, 2012.

Jacobsen, Rowan. *Fruitless Fall: The Collapse of the Honeybee and the Coming Agricultural Crisis.* Bloomsbury USA, 2009.

Kimbrell, Andrew. *Fatal Harvest: The Tragedy of Industrial Agriculture.* Foundation for Deep Ecology, 2002.

L. L. Langstroth. *Langstroth on the Hive and the Honey-Bee: A Bee Keeper's Manual.* Hopkins, Bridgman & Company, 1853. [online]. [cited August 27, 2012]. gutenberg.org/files/24583/24583-h/24583-h.htm.

Pollan, Michael. *The Omnivore's Dilemma: A Natural History of Four Meals.* Penguin, 2006.

—. *The Botany of Desire: A Plant's-Eye View of the World.* Random, 2001.

Root, Amos Ives, Ann Harman, Dr. Hachiro Shimanuki and Kim Flottum. *The ABC & XYZ of Bee Culture: An Encyclopedia Pertaining to the Scientific and Practical Culture of Honey Bees*, 41st ed. A.I. Root, 2007.

Seeley, Thomas D. *Honeybee Democracy*. Princeton, 2010.
Steiner, Rudolph. *Bees*. Steiner Press, 1998.
Stiglitz, Dean and Laurie Herboldsheimer. *The Complete Idiot's Guide to Beekeeping*. Alpha, 2010.

Movies
Dirt! The Movie, 1 hour 21 minutes, New Video, 2009.
Food, Inc., 1 hour 34 minutes, Magnolia, 2008.
King Corn: You Are What You Eat, 3 hours 1 minute, New Video, 2007.
Nature: Silence of the Bees, 118 minutes, Questar, 2008.
Nicotine Bees, 55 minutes, Nicotine Bees, 2010.
Queen of the Sun: What Are the Bees Telling Us?, 82 minutes, Music Box Films and Collective Eye, Inc., 2012.
Vanishing of the Bees, 87 minutes, True Mind, 2011.

Websites: General
The Barefoot Beekeeper. [online]. [cited May 21, 2012]. biobees.com.
Bushfarms Beekeeping. [online]. [cited May 21, 2012]. bushfarms.com/bees.htm.
Gold Star Honeybees®. [online]. [cited May 21, 2012]. goldstarhoneybees.com.
John's Beekeeping Notebook. [online]. [cited May 21, 2012]. outdoorplace.org/bee keeping/index.htm.
Zach Huang. *Gallery* [of bee photos]. [online]. [cited September 4, 2012]. cyberbee.net/gallery/index.php/

Websites: Bee Magazines
American Bee Journal. [online]. [cited May 21, 2012]. americanbeejournal.com.
Bee Culture: The Magazine of American Beekeeping. [online]. [cited May 21, 2012]. beeculture.com.

Websites: Beekeeping Associations
American Beekeeping Federation. [online]. [cited May 22, 2012]. abfnet.org.
Canadian Honey Council. [online]. [cited May 22, 2012]. honeycouncil.ca.
EAS — Eastern Apicultural Society. [online]. [cited May 22, 2012]. easternapiculture .org.
HAS — Heartland Apicultural Society. [online]. [cited May 22, 2012]. heartlandbees .com.
[US] *National Honey Board*. [online]. [cited May 22, 2012]. honey.com.
WAS — Western Apicultural Society. [online]. [cited May 22, 2012]. groups.ucanr .org/WAS.

Websites: Bee Research

Many universities are great bee research sites. I have mentioned a few below, but be sure to check institutions near you.

EXtension.org. *Bee Health*. [online]. [cited September 4, 2012]. eXtension.org/bee _health.

Mid-Atlantic Apiculture Research and Extension Consortium (MAAREC). [online]. [cited September 8, 2012]. agdev.anr.udel.edu/maarec/.

Penn State College of Agricultural Sciences, Entomology. *Honey Bee and Pollinator Research*. [online]. [cited September 4, 2012]. ento.psu.edu/directory/research -issues/honey-bee-and-pollinator-research.

University of Georgia College of Agricultural and Environmental Sciences. *Honey Bee Program*. [online]. [cited September 4, 2012]. ent.uga.edu/bees/.

University of Minnesota. *Bee Lab*. [online]. [cited September 4, 2012]. beelab.umn .edu/.

USDA Agricultural Research Service. *USDA-ARS Bee Labs*. [online]. [cited September 4, 2012]. eXtension.org/pages/21736/usda-ars-bee-labs.

How to find a Beekeeper Near You

Bee Culture Magazine. *Who's Who in North American Beekeeping*. [online]. [cited August 30, 2012]. beeculture.com/content/whoswho/.

Index

About the Author

So…just what did bees do before beekeepers?

Searching for an answer to what seemed a simple question, Christy Hemenway launched her own investigation into what was really behind the growing problems with honeybees. She soon came to the conclusion that with honeybees, less is more—in other words, less human manipulation is better for the honeybee.

This insight led Christy to found Gold Star Honeybees® in 2007—to advance a low-tech, natural beekeeping system known as the top bar hive. The most important feature of a top bar hive is that it allows the bees to make their beeswax honeycombs in accord with their own natural systems, in a non-toxic hive environment. A Gold Star top bar hive is clean and green and supports the making of natural beeswax—beeswax made by bees, for bees!

In her TEDxDirigo talk in 2011, "Making the Connection—Honeybees, Food and You," Christy highlighted the important connection between our agricultural system, honeybees and health. Spreading this same message through her bee-evangelist speaker persona The Bee Lady, Christy advocates and agitates for more organic food, less industrial agriculture and of course more natural, less invasive beekeeping.

Visit Gold Star Honeybees® on the web at goldstarhoneybees.com.

Find us on Facebook at facebook.com/goldstarhoneybees.

**Christy Hemenway—
The Bee Lady.**
Credit: Chris Hendricks.

If you have enjoyed *The Thinking Beekeeper,* you might also enjoy other

BOOKS TO BUILD A NEW SOCIETY

Our books provide positive solutions for people who
want to make a difference. We specialize in:

**Sustainable Living • Green Building • Peak Oil
Renewable Energy • Environment & Economy
Natural Building & Appropriate Technology
Progressive Leadership • Resistance and Community
Educational & Parenting Resources**

New Society Publishers

ENVIRONMENTAL BENEFITS STATEMENT

New Society Publishers has chosen to produce this book on recycled paper made
with **100% post consumer waste,** processed chlorine free, and old growth free.

For every 5,000 books printed, New Society saves the following resources:[1]

27	Trees
2,470	Pounds of Solid Waste
2,718	Gallons of Water
3,545	Kilowatt Hours of Electricity
4,490	Pounds of Greenhouse Gases
19	Pounds of HAPs, VOCs, and AOX Combined
7	Cubic Yards of Landfill Space

[1]Environmental benefits are calculated based on research done by the Environmental Defense Fund
and other members of the Paper Task Force who study the environmental impacts of the paper
industry.

For a full list of NSP's titles, please call 1-800-567-6772 *or check out our website* at:

www.newsociety.com